T0269297

LONDON MATHEMATICAL SOCIETY LECTURE NOTE SERIES

Managing Editor: Professor J.W.S. Cassels, Department of Pure Mathematics and Mathematical Statistics, University of Cambridge, 16 Mill Lane, Cambridge CB2 1SB, England

The books in the series listed below are available from booksellers, or, in case of difficulty, from Cambridge University Press.

34 Representation theory of Lie groups, M.F. ATIYAH *et al*
36 Homological group theory, C.T.C. WALL (ed)
39 Affine sets and affine groups, D.G. NORTHCOTT
46 p-adic analysis: a short course on recent work, N. KOBLITZ
49 Finite geometries and designs, P. CAMERON, J.W.P. HIRSCHFELD & D.R. HUGHES (eds)
50 Commutator calculus and groups of homotopy classes, H.J. BAUES
57 Techniques of geometric topology, R.A. FENN
59 Applicable differential geometry, M. CRAMPIN & F.A.E. PIRANI
66 Several complex variables and complex manifolds II, M.J. FIELD
69 Representation theory, I.M. GELFAND *et al*
74 Symmetric designs: an algebraic approach, E.S. LANDER
76 Spectral theory of linear differential operators and comparison algebras, H.O. CORDES
77 Isolated singular points on complete intersections, E.J.N. LOOIJENGA
79 Probability, statistics and analysis, J.F.C. KINGMAN & G.E.H. REUTER (eds)
80 Introduction to the representation theory of compact and locally compact groups, A. ROBERT
81 Skew fields, P.K. DRAXL
82 Surveys in combinatorics, E.K. LLOYD (ed)
83 Homogeneous structures on Riemannian manifolds, F. TRICERRI & L. VANHECKE
86 Topological topics, I.M. JAMES (ed)
87 Surveys in set theory, A.R.D. MATHIAS (ed)
88 FPF ring theory, C. FAITH & S. PAGE
89 An F-space sampler, N.J. KALTON, N.T. PECK & J.W. ROBERTS
90 Polytopes and symmetry, S.A. ROBERTSON
91 Classgroups of group rings, M.J. TAYLOR
92 Representation of rings over skew fields, A.H. SCHOFIELD
93 Aspects of topology, I.M. JAMES & E.H. KRONHEIMER (eds)
94 Representations of general linear groups, G.D. JAMES
95 Low-dimensional topology 1982, R.A. FENN (ed)
96 Diophantine equations over function fields, R.C. MASON
97 Varieties of constructive mathematics, D.S. BRIDGES & F. RICHMAN
98 Localization in Noetherian rings, A.V. JATEGAONKAR
99 Methods of differential geometry in algebraic topology, M. KAROUBI & C. LERUSTE
100 Stopping time techniques for analysts and probabilists, L. EGGHE
101 Groups and geometry, ROGER C. LYNDON
103 Surveys in combinatorics 1985, I. ANDERSON (ed)
104 Elliptic structures on 3-manifolds, C.B. THOMAS
105 A local spectral theory for closed operators, I. ERDELYI & WANG SHENGWANG
106 Syzygies, E.G. EVANS & P. GRIFFITH
107 Compactification of Siegel moduli schemes, C-L. CHAI
108 Some topics in graph theory, H.P. YAP
109 Diophantine analysis, J. LOXTON & A. VAN DER POORTEN (eds)
110 An introduction to surreal numbers, H. GONSHOR
111 Analytical and geometric aspects of hyperbolic space, D.B.A. EPSTEIN (ed)
113 Lectures on the asymptotic theory of ideals, D. REES
114 Lectures on Bochner-Riesz means, K.M. DAVIS & Y-C. CHANG
115 An introduction to independence for analysts, H.G. DALES & W.H. WOODIN
116 Representations of algebras, P.J. WEBB (ed)
117 Homotopy theory, E. REES & J.D.S. JONES (eds)
118 Skew linear groups, M. SHIRVANI & B. WEHRFRITZ
119 Triangulated categories in the representation theory of finite-dimensional algebras, D. HAPPEL
121 Proceedings of *Groups - St Andrews 1985*, E. ROBERTSON & C. CAMPBELL (eds)
122 Non-classical continuum mechanics, R.J. KNOPS & A.A. LACEY (eds)
124 Lie groupoids and Lie algebroids in differential geometry, K. MACKENZIE
125 Commutator theory for congruence modular varieties, R. FREESE & R. MCKENZIE
126 Van der Corput's method of exponential sums, S.W. GRAHAM & G. KOLESNIK
127 New directions in dynamical systems, T.J. BEDFORD & J.W. SWIFT (eds)

128 Descriptive set theory and the structure of sets of uniqueness, A.S. KECHRIS & A. LOUVEAU
129 The subgroup structure of the finite classical groups, P.B. KLEIDMAN & M.W.LIEBECK
130 Model theory and modules, M. PREST
131 Algebraic, extremal & metric combinatorics, M-M. DEZA, P. FRANKL & I.G. ROSENBERG (eds)
132 Whitehead groups of finite groups, ROBERT OLIVER
133 Linear algebraic monoids, MOHAN S. PUTCHA
134 Number theory and dynamical systems, M. DODSON & J. VICKERS (eds)
135 Operator algebras and applications, 1, D. EVANS & M. TAKESAKI (eds)
136 Operator algebras and applications, 2, D. EVANS & M. TAKESAKI (eds)
137 Analysis at Urbana, I, E. BERKSON, T. PECK, & J. UHL (eds)
138 Analysis at Urbana, II, E. BERKSON, T. PECK, & J. UHL (eds)
139 Advances in homotopy theory, S. SALAMON, B. STEER & W. SUTHERLAND (eds)
140 Geometric aspects of Banach spaces, E.M. PEINADOR and A. RODES (eds)
141 Surveys in combinatorics 1989, J. SIEMONS (ed)
142 The geometry of jet bundles, D.J. SAUNDERS
143 The ergodic theory of discrete groups, PETER J. NICHOLLS
144 Introduction to uniform spaces, I.M. JAMES
145 Homological questions in local algebra, JAN R. STROOKER
146 Cohen-Macaulay modules over Cohen-Macaulay rings, Y. YOSHINO
147 Continuous and discrete modules, S.H. MOHAMED & B.J. MÜLLER
148 Helices and vector bundles, A.N. RUDAKOV et al
149 Solitons, nonlinear evolution equations and inverse scattering, M.J. ABLOWITZ
     & P.A. CLARKSON
150 Geometry of low-dimensional manifolds 1, S. DONALDSON & C.B. THOMAS (eds)
151 Geometry of low-dimensional manifolds 2, S. DONALDSON & C.B. THOMAS (eds)
152 Oligomorphic permutation groups, P. CAMERON
153 L-functions and arithmetic, J. COATES & M.J. TAYLOR (eds)
154 Number theory and cryptography, J. LOXTON (ed)
155 Classification theories of polarized varieties, TAKAO FUJITA
156 Twistors in mathematics and physics, T.N. BAILEY & R.J. BASTON (eds)
157 Analytic pro-p groups, J.D. DIXON, M.P.F. DU SAUTOY, A. MANN & D. SEGAL
158 Geometry of Banach spaces, P.F.X. MÜLLER & W. SCHACHERMAYER (eds)
159 Groups St Andrews 1989 volume 1, C.M. CAMPBELL & E.F. ROBERTSON (eds)
160 Groups St Andrews 1989 volume 2, C.M. CAMPBELL & E.F. ROBERTSON (eds)
161 Lectures on block theory, BURKHARD KÜLSHAMMER
162 Harmonic analysis and representation theory for groups acting on homogeneous trees,
     A. FIGA-TALAMANCA & C. NEBBIA
163 Topics in varieties of group representations, S.M. VOVSI
164 Quasi-symmetric designs, M.S. SHRIKANDE & S.S. SANE
165 Groups, combinatorics & geometry, M.W. LIEBECK & J. SAXL (eds)
166 Surveys in combinatorics, 1991, A.D. KEEDWELL (ed)
167 Stochastic analysis, M.T. BARLOW & N.H. BINGHAM (eds)
168 Representations of algebras, H. TACHIKAWA & S. BRENNER (eds)
169 Boolean function complexity, M.S. PATERSON (ed)
170 Manifolds with singularities and the Adams-Novikov spectral sequence, B. BOTVINNIK
172 Algebraic varieties, GEORGE R. KEMPF
173 Discrete groups and geometry, W.J. HARVEY & C. MACLACHLAN (eds)
174 Lectures on mechanics, J.E. MARSDEN
175 Adams memorial symposium on algebraic topology 1, N. RAY & G. WALKER (eds)
176 Adams memorial symposium on algebraic topology 2, N. RAY & G. WALKER (eds)
177 Applications of categories in computer science, M.P. FOURMAN, P.T. JOHNSTONE,
     & A.M. PITTS (eds)
178 Lower K- and L-theory, A. RANICKI
179 Complex projective geometry, G. ELLINGSRUD, C. PESKINE, G. SACCHIERO
     & S.A. STRØMME (eds)
180 Lectures on ergodic theory and Pesin theory on compact manifolds, M. POLLICOTT
181 Geometric group theory I, G.A. NIBLO & M.A. ROLLER (eds)
182 Geometric group theory II, G.A. NIBLO & M.A. ROLLER (eds)
184 Arithmetical functions, W. SCHWARZ & J. SPILKER
185 Representations of solvable groups, O. MANZ & T.R. WOLF
186 Complexity: knots, colourings and counting, D.J.A. WELSH
187 Surveys in combinatorics, 1993, K. WALKER (ed)
190 Polynomial invariants of finite groups, D.J. BENSON

London Mathematical Society Lecture Note Series. 172

Algebraic Varieties

George R. Kempf
Johns Hopkins University

CAMBRIDGE
UNIVERSITY PRESS

CAMBRIDGE UNIVERSITY PRESS
Cambridge, New York, Melbourne, Madrid, Cape Town, Singapore,
São Paulo, Delhi, Tokyo, Mexico City

Cambridge University Press
The Edinburgh Building, Cambridge CB2 8RU, UK

Published in the United States of America by Cambridge University Press, New York

www.cambridge.org
Information on this title: www.cambridge.org/9780521426138

First published 1993

A catalogue record for this publication is available from the British Library

ISBN 978-0-521-42613-8 Paperback

Contents

Introduction   ix

1 Algebraic varieties: definition and existence   1

1.1   Spaces with functions   1
1.2   Varieties   2
1.3   The existence of affine varieties   4
1.4   The nullstellensatz   5
1.5   The rest of the proof of existence of affine
       varieties / subvarieties   8
1.6 $\mathbf{A}^n$ and $\mathbf{P}^n$ 10
1.7   Determinantal varieties   11

2 The preparation lemma and some consequences   13

2.1   The lemma   13
2.2   The Hilbert basis theorem   15
2.3   Irreducible components   16
2.4   Affine and finite morphisms   18
2.5   Dimension   20
2.6   Hypersurfaces and the principal ideal theorem   21

3 Products; separated and complete varieties   25

3.1   Products   25

3.2   Products of projective varieties   27
3.3   Graphs of morphisms and separatedness   28
3.4   Algebraic groups   30
3.5   Cones and projective varieties   31
3.6   A little more dimension theory   32
3.7   Complete varieties   33
3.8   Chow's lemma   34
3.9   The group law on an elliptic curve   35
3.10   Blown up $\mathbf{A}^n$ at the origin   36

4 Sheaves   38

4.1   The definition of presheaves and sheaves   38
4.2   The construction of sheaves   42
4.3   Abelian sheaves and flabby sheaves   46
4.4   Direct limits of sheaves   50

5 Sheaves in algebraic geometry   54

5.1   Sheaves of rings and modules   54
5.2   Quasi-coherent sheaves on affine varieties   56
5.3   Coherent sheaves   58
5.4   Quasi-coherent sheaves on projective varieties   61
5.5   Invertible sheaves   62
5.6   Operations on sheaves that change spaces   65
5.7   Morphisms to projective space and affine morphisms   68

6 Smooth varieties and morphisms   70

6.1   The Zariski cotangent space and smoothness   70
6.2   Tangent cones   72
6.3   The sheaf of differentials   75
6.4   Morphisms   80
6.5   The construction of affine morphisms and normalization   82
6.6   Bertini's theorem   83

7 Curves   85

7.1   Introduction to curves   85
7.2   Valuation criterions   87
7.3   The construction of all smooth curves   88
7.4   Coherent sheaves on smooth curves   90
7.5   Morphisms between smooth complete curves   92

7.6   Special morphisms between curves   94
7.7   Principal parts and the Cousin problem   96

8 Cohomology and the Riemann–Roch theorem   98

8.1   The definition of cohomology   98
8.2   Cohomology of affines   100
8.3   Higher direct images   102
8.4   Beginning the study of the cohomology of curves   104
8.5   The Riemann–Roch theorem   106
8.6   First applications of the Riemann–Roch theorem   108
8.7   Residues and the trace homomorphism   110

9 General cohomology   113

9.1   The cohomology of $\mathbf{A}^n - \{0\}$ and $\mathbf{P}^n$ 113
9.2   Čech cohomology and the Künneth formula   114
9.3   Cohomology of projective varieties   116
9.4   The direct images of flat sheaves   118
9.5   Families of cohomology groups   120

10 Applications   124

10.1   Embedding in projective space   124
10.2   Cohomological characterization of affine varieties   125
10.3   Computing the genus of a plane curve and
         Bezout's theorem   126
10.4   Elliptic curves   128
10.5   Locally free coherent sheaves on $\mathbf{P}^1$ 129
10.6   Regularity in codimension one   130
10.7   One dimensional algebraic groups   131
10.8   Correspondences   132
10.9   The Reimann–Roch theorem for surfaces   139

Appendix   139

A.1   Localization   141
A.2   Direct limits   143
A.3   Eigenvectors   144

Bibliography   146
Glossary of notation   149
Index   155

Introduction

Algebraic geometry is a mixture of the ideas of two Mediterranean cultures. It is the superposition of the Arab science of the lightning calculation of the solutions of equations over the Greek art of position and shape. This tapestry was originally woven on European soil and is still being refined under the influence of international fashion. Algebraic geometry studies the delicate balance between the geometrically plausible and the algebraically possible. Whenever one side of this mathematical teeter-totter outweighs the other, one immediately loses interest and runs off in search of a more exciting amusement.

In this book we present from a modern point of view the basic theory of algebraic varieties and their coherent cohomology. The local part of the study includes dimension and smoothness. I have tried to keep the commutative algebra down to minimum while putting the geometry close to the algebra as part of the exposition.

The basic tools in algebraic geometry are sheaves and their cohomology. This material is presented from the beginning. I have included the basic discussion of curves to illustrate the theory.

To proceed further in algebraic geometry one needs to learn scheme language. This transition should be easy for the reader of this book. One needs only to drop the assumption that the structure sheaf consists of functions. In writing this book I missed having generic points of subvarieties and closed subschemes. But the main battle was to teach the reader to think globally in sheaf-theoretic language.

There are many good presentations of more advanced material. I personally recommend Mumford's lectures on curves on a surface and Grothendieck's *Elements of Algebraic Geometry* – or for theory of curves and their Jacobians my *Abelian Integrals*.

I have enclosed an appendix on localization and direct limits of sets. Furthermore, all rings are assumed to be commutative with identity.

I thank Mr. S. T. Soh for many corrections.

1

Algebraic varieties:
definition and existence

In this chapter we meet the category of algebraic varieties. We will
give their definitions and discuss the subsequent question of their existence. We begin the discussion with the larger category of spaces with
functions.

1.1 Spaces with functions

Let k be a fixed field. *A space with functions* is a topological space X
together with the assignment to each open subset U of X of a k-algebra
$k[U]$ of k-valued functions on U, which we say consists of all *regular*
functions on U, satisfying the properties (a) and (b) below.
The conditions are:

(a)   Let U be the union $\bigcup U_\alpha$ of a family of open subsets. Let f be a
   k-valued function on U. Then f is regular on U iff the restriction
   $f|_{U_\alpha}$ is regular on each U_α.

(b)   Let f be a regular function on an open subset U. Then $D(f) \equiv$
   $\{u \in U | f(u) \neq 0\}$ is open and $\frac{1}{f}$ is regular on $D(f)$.

Thus (a) says that a function is regular iff it is locally regular. Also
we may add, subtract, multiply and divide regular functions whenever
it is reasonable and constant functions are regular.

A first example of a space with functions is:

Example. If $k = \mathbf{R}$ or $\mathbf{C}$ and X is any topological space, then X has a natural structure of a space with functions. Just take $k[U]$ as the set of all continuous functions: $U \to k$.

In algebraic geometry $k[U]$ is commonly denoted by $\mathcal{O}_X(U)$. In general we shall denote a space with functions by the topological space X with the rings of locally regular functions being understood.

A second example of a space with functions is an *open subspace* of a space with functions. Let V be an open subset of a space with functions X. Give V the subspace topology. Then if U is an open subset of V then a regular function on U for the structure of V is simply a regular function on U for the structure on X; i.e., $\mathcal{O}_V(U) = \mathcal{O}_X(U)$.

A *morphism* $f : X \to Y$ between two spaces with functions is a continuous mapping which pulls-back regular functions into regular functions; i.e., if $g(v)$ is a regular function on an open subset V of Y then $f^\star(g)(u) \equiv g(f(u))$ is a regular function on the open subset $f^{-1}(V)$ of X . Thus pulling-back by f defines a k-algebra homomorphism $f^\star : k[V] \to k[f^{-1}V]$ for each open subset V of Y. An *isomorphism* is a bijective mapping f such that both f and f^{-1} are morphisms.

Exercise 1.1.1. In the first example show that any continuous mapping $X_1 \to X_2$ is a morphism.

Exercise 1.1.2. Prove that the identity of a space with functions is a morphism and the composition of morphisms is a morphism.

Exercise 1.1.3. Let U be an open subspace of a space with functions X. Then show that the inclusion $i : U \to X$ is a morphism and if Y is another space with functions and $g : Y \to U$ is a mapping, then g is a morphism if and only if $i \circ g : Y \to X$ is.

Exercise 1.1.4. Let $f : X \to Y$ be a mapping between spaces with functions. Given open covers $X = \bigcup U_\alpha$ and $Y = \bigcup V_\alpha$ such that $f(U_\alpha) \subseteq V_\alpha$, f is a morphism if and only if each $f_\alpha : U_\alpha \to V_\alpha$ is.

1.2 Varieties

We shall henceforth assume that the field k is algebraically closed.

Let X and Y be spaces with functions. The global effect of pull-back defines a mapping

$$\star : \mathrm{Morphism}(X, Y) \to k\text{-Alg-Hom}(k[Y], k[X])$$

which sends a morphism f to $f^\star$.

An *affine variety* Y is a space with functions such that $\star$ is bijective for every X and $k[Y]$ is a finitely generated k-algebra. An (algebraic) *variety* X is a space with functions X which has a finite open covering $U_1, \ldots, U_n$ where each U_i is affine. A *morphism* of varieties is just a morphism of spaces with functions.

In this section we will give the first examples of varieties: the *affine line* $\mathbf{A}^1$ and the *projective line* $\mathbf{P}^1$.

As a set $\mathbf{A}^1 = \{(x)\}$ is just k. The closed subsets of $\mathbf{A}^1$ are the whole $\mathbf{A}^1$ and the finite subsets. This gives $\mathbf{A}^1$ its topology. Let U be an open subset of $\mathbf{A}^1$. If U is empty, $k[U] = \{0\}$. Otherwise if $U = \mathbf{A}^1 - \{x_1, \ldots, x_n\}$ then $k[U]$ consists of the rational functions $g(x)$ in the coordinate x such that g has no poles anywhere in U; i.e. $g(x) = \frac{p(x)}{\prod (x - x_i)^{m_i}}$ where p is a polynomial and the m_i non-negative integers. In particular $k[\mathbf{A}^1] = k[X]$ is a polynomial ring in one variable. We leave the details of checking that $\mathbf{A}^1$ is a space with functions as a very instructive exercise. Here we will check that $\mathbf{A}^1$ is affine.

Let Y be any space with functions. A mapping $f : Y \to \mathbf{A}^1 = k$ is just a k-valued function.

Claim. f is a morphism if and only if f is regular on Y.

Proof. "Only if" is obvious because the function $f = f^*$(coordinate function x). Conversely, we need to see that if f is regular then f is a morphism; i.e., $f^{-1}(\{x_1, \ldots, x_n\})$ is closed but $f^{-1}(\{x_1, \ldots, x_n\})$ is the complement of the open subset $D(\prod (f - x_i))$ of Y. To show that f pulls-back regular functions, just note $f^*g = \frac{p(f(y))}{\prod (f(y) - x_i)^{m_i}}$ is regular off $f^{-1}(\{x_1, \ldots, x_n\})$ where g is as before.

To finish we have $\star$: Morphism$(Y, \mathbf{A}^1) \approx k[Y] \approx k$-Alg-Hom $(k[\mathbf{A}^1], k[Y])$. This gives the proof that $\mathbf{A}^1$ is affine. $\qquad\square$

As a set $\mathbf{P}^1 = k \amalg \{\infty\}$ where ∞ is a symbol. The non-trivial closed subsets of $\mathbf{P}^1$ are again finite. A regular function on $\mathbf{P}^1 - \{x_1, \ldots, x_n\}$ is a rational function of x which *has no poles* except at $x_1, \ldots, x_n$ where $g(x)$ has no pole at $\infty \equiv g(\frac{1}{y})$ has no pole at $y = 0$. To see that $\mathbf{P}^1$ is a variety it has an open covering $(\mathbf{P}^1 - \{\infty\}) \cup (\mathbf{P}^1 - \{0\})$ by two affine lines. Here $\mathbf{A}^1 \approx \mathbf{P}^1 - \{\infty\}$ sends x to x and $\mathbf{A}^1 \approx \mathbf{P}^1 - \{0\}$ sends x to $1/x$. An interesting feature of this example is:

Exercise 1.2.1. $k[\mathbf{P}^1] = k$.

Exercise 1.2.2. Prove in detail that $\mathbf{A}^1$ is a space with functions.

Exercise 1.2.3. Why is $\mathbf{P}^1$ not affine?

Exercise 1.2.4. Let X and Y be two affine varieties. Show that X is isomorphic to Y iff the k-algebras $k[X]$ and $k[Y]$ are isomorphic.

1.3 The existence of affine varieties

If you did Exercise 1.2.2 you may have already noticed that it is not completely trivial to check that a space with functions is affine. The main work of this chapter is to explicitly construct all affine varieties. The main result is

Theorem 1.3.1. *If A is a finitely generated k-algebra with no nilpotents, then there is a canonically constructed affine variety* Spec A *with a natural isomorphism*

$$A \approx k[\text{Spec } A].$$

Recall that A has no nilpotents if $a^n = 0$ for some $n > 0$ implies $a = 0$ in A. Clearly a ring of k-valued functions has no nilpotents. Therefore the hypothesis of the theorem is necessary for A to be k[affine variety]. So the theorem constructs all affine varieties up to isomorphism (see Exercise 1.2.4).

In this section we will define Spec A as a space with functions together with the homomorphism $\phi : A \to k[\text{Spec } A]$. The rest of this chapter is devoted to the proof that

($\star$) Spec A is affine and

($\star\star$) ϕ is an isomorphism.

As a set Spec $A = k$-Alg-Hom (A, k). We have a natural k-algebra homomorphism $\phi : A \to \{k$-valued functions on Spec $A\}$. Just let $\phi(a)(x) \equiv x(a)$ where x is a point of Spec A.

Let I be a subset of A. Define zeroes$(I) = \{x \in \text{Spec } A | i(x) = 0$ for all $i \in I\}$.

Claim. The subsets $\{\text{zeroes}(I)\}_{I \subset A}$ are the closed subsets of a topology of Spec A.

Proof.

(a) Spec A = zeroes($\{0\}$) and $\emptyset$ = zeroes($\{1\}$).

(b) zeroes$(I_1 \cdot I_2)$ = zeroes$(I_1) \bigcup$ zeroes(I_2).

(c) zeroes$(\bigcup I_i) = \bigcap$ zeroes(I_i). Why?

$\square$

If U is an open subset of $\mathrm{Spec}A$, a regular function on U is a k-valued function f on U such that there is an open covering $U = \bigcup U_i$ such that each $f|_{U_i}$ has the form $\frac{\phi(a_i)(x)}{\phi(b_i)(x)}$ where the denominator $\phi(b_i)(x)$ never vanishes on U_i. Clearly $k[U]$ is a k-algebra and condition (a) is easily verified. For condition (b) note that $D(f) = \bigcup(U_i - \mathrm{zeroes}(a_i) \cap U_i)$ is open and $1/f \left(= \frac{\phi(b_i)x}{\phi(a_i)x} \right)$ is regular on $D(f)$. As $\phi(a)(x) = \frac{\phi(a)(x)}{\phi(1)(x)}$ the image of ϕ is contained in $k[\mathrm{Spec}A]$, that is $\phi : A \to k[\mathrm{Spec}\ A]$.

Exercise 1.3.2. Check the above details.

Exercise 1.3.3. Show that, for any ring A, the set of nilpotent elements $\sqrt{0} = \{a \in A | a^n = 0 \text{ for some } n > 0\}$ is an ideal. Also check that $A = \sqrt{0} \iff A = \{0\}$.

Exercise 1.3.4. Set-theoretically, what is Spec A when A is the polynomial ring $k[X_1, \ldots, X_n]$? Same if $A = k[X_1, \ldots, X_n]/(f_1, \ldots, f_n, \ldots,)$ for some polynomials f_i.

Exercise 1.3.5. If A is a finitely generated ring with no nilpotents, let a be an element of A. Consider the homomorphism $\phi : A \to A_{(a)}$, where $A_{(a)}$ is a finitely generated k-algebra with no nilpotents (Why?). Show that ϕ induces an isomorphism $\mathrm{Spec}(A_{(a)}) \xrightarrow{\approx} D(a) \subset \mathrm{Spec}\ A$ where $D(a)$ is given the open subspace structure as a space with functions.

Exercise 1.3.6. Show that the collection $\{D(a)\}_{a \in A}$ are a basis for the topology of Spec A and $D(a_1) \cap D(a_2) = D(a_1 \cdot a_2)$.

1.4 The nullstellensatz

The objective of this section is to prove Hilbert's nullstellensatz which says that Spec A has enough points so that

$(**)_1$ $\qquad\qquad\qquad \phi : A \to k[\mathrm{Spec}\ A]$

is injective.

We will begin with a lemma of E. Noether whose proof will be presented in the second chapter.

Lemma 1.4.1. *Let A be a non-zero finitely generated k-algebra. Then we have an injection $B \subset A$ where B is a polynomial ring $k[X_1, \ldots, X_d]$ such that A is a B-module of finite type.*

Next we have

Lemma 1.4.2. *Let $k \subset B \subset A$ be k-algebras such that A is a B-module of finite type. Then composition gives a surjection*

$$k\text{-Alg-Hom}(A, k) \to k\text{-Alg-Hom}(B, k).$$

Proof. Let $\psi : B \to k$ be a k-algebra homomorphism. Let m be the kernel of ψ. Consider the ideal mA in A. If $mA \neq A$, we can find a maximal ideal n of A such that $n \supset mA$. Then A/n is a finite field extension of $B/m = k$. Thus $A/n = k$ as k is algebraically closed and the quotient mapping $A \to k$ is an extension of ψ.

It remains to show that $mA = A$ is impossible. Let $A = B^r/R$ as a B-module where R is a B-submodule of B^r where r is finite. Let $e_1, \ldots, e_r$ be the unit vectors in B^r. Then for $1 \leq i \leq r$, $e_i \in mB^r + R$, i.e., $e_j = \sum_k m_j^k e_k + r_j$ where the r_j are in R and the m_j^k are in m. So if $(b_j^k) \equiv 1_r - (m_j^k)$ then $r_j = \sum b_j^k e_k$ is in R. By Cramer's rule $\det(b_j^k)e_i$ is a linear combination of the r_j with coefficients in B. Thus $\det(b_j^k)B^r \subseteq R$ and hence $\det(b_j^k) \cdot A = 0$, or, what is the same, $\det(b_j^k) = 0$. On the other hand $\det(b_j^k) = 1((m))$. As $1 \neq 0((m))$ we have a contradiction. $\square$

With these lemmas the proof of the nullstellensatz is easy. Let a be a non-zero element of A. Consider the localization $A_{(a)}$. This is non-zero as $1/1 = 0/1 \Leftrightarrow a^n = a^n \cdot 1 = a^n \cdot 0 = 0$ for $n > 0$ and A has no nilpotents. Furthermore $A_{(a)}$ is generated by A and $1/a$ hence is finitely generated. Thus by Lemma 1.4.1 we have an inclusion $k[X_1, \ldots, X_r] \subset A_{(a)}$.

We need to find a point x of Spec A such that $\phi(a)(x) \equiv x(a)$ is non-zero where $x : A \to k$ is a k-algebra homomorphism. Let $z : k[X_1, \ldots, X_r] \to k$ be any k-algebra homomorphism. By Lemma 1.4.2 we may lift z to a k-algebra homomorphism $y : A_{(a)} \to k$. Let x be the composition $A \to A_{(a)} \xrightarrow{y} k$. Then $1 = x(1) = y(a/a) = y(a/1)y(1/a) = x(a)y(1/a)$. Therefore $x(a) \neq 0$. This proves the nullstellensatz.

Remark. The argument with Cramer's rule in the last part of Lemma 1.4.2 can be generalized to prove

Lemma 1.4.3. (Nakayama.) *Let M be a finitely generated module over a ring A. Assume that there is an ideal I of A such that $M = I \cdot M$. Then there is an element a of $1 + I$ such that $aM = \{0\}$.*

Exercise 1.4.4. Prove Nakayama's lemma.

Henceforth we shall identify A with a ring of functions of Spec A. Next we will give a reformulation of the nullstellensatz which resembles Hilbert's original statement.

Theorem 1.4.5. (Nullstellensatz.) *If I is an ideal of A, $\{a \in A | a(x) = 0$ for all x in zeroes$(I)\} = \sqrt{I}$ where $\sqrt{I} = \{a \in A | a^n \in I$ for some $n > 0\}$.*

Proof. One checks that $\sqrt{I}$ is an ideal. Then $A' = A/\sqrt{I}$ is a finitely generated ring which has no nilpotents. Let b be an element of $A - \sqrt{I}$. Then $b' = b + \sqrt{I}$ is a non-zero element of A'. By $(**)_1$, for A' we have a homomorphism $x' : A' \to k$ such that $x'(b') \neq 0$. Let $x : A \to A' \xrightarrow{x'} k$ be the composition. Then by construction x is a point of zeroes(I) and $b(x) \neq 0$. This proves the desired inclusion "$\subseteq$". The reverse inclusion is trivial because $b^n(x) = 0 \Leftrightarrow b(x) = 0$ if $n > 0$. $\qquad\square$

Some special cases will be useful.

Corollary 1.4.6. *Given a subset J of A then* zeroes $(J) = \emptyset \Leftrightarrow 1 = \sum_{\text{finite}} a_k j_k$ *where $a_k \in A$ and $j_k \in J$.*

Proof. Apply Theorem 1.4.5 to the ideal $I = JA$. Hence if zeroes$(J) = \emptyset$, then zeroes$(I) = \emptyset$ and we have $A = \sqrt{I}$; i.e., $1 = 1^n \in JA$ which is the second statement. The converse is obvious. $\qquad\square$

Corollary 1.4.7. *Let f_i be elements of A and n_i be positive integers.*

$$\text{Spec}(A) = \bigcup D(f_i) \Leftrightarrow 1 = \sum_{\text{finite}} a_i f_i^{n_i}.$$

Proof. The complement of the open subset $\bigcup D(f_i)$ is zeroes$\{(f_i^{n_i})\}$. Thus this corollary follows from the last. $\qquad\square$

This result has an interesting topological consequence.

Corollary 1.4.8. *A variety is quasi-compact.*

Proof. As a variety is a finite union of open affines, it suffices to prove this for the affine variety Spec A. Now the $D(f_i)$ are a basis for the topology of Spec A. If Spec $A = \bigcup D(f_i)$ is an open cover by this then

$1 = \sum a_i f_i$ for f_i in a finite set I of indexes. Then Spec $A = \bigcup_{i \in I} D(f_i)$ is a finite subcover. $\square$

Exercise 1.4.9. Prove that a point of a variety is a closed subset. (Hint: reduce to the affine case.)

1.5. The rest of the proof of existence of affine varieties / subvarieties

We will first show

$(**)_2$ The subring $A \subset k[\text{Spec } A]$ is all of $k[\text{Spec } A]$.

Let $f(x)$ be a function in $k[\text{Spec } A]$. We need to see that f is in A. By definition we have an open cover Spec $A = \bigcup U_\alpha$ by open subsets such that $f(u) = \frac{g_\alpha(u)}{h_\alpha(u)}$ when u in U_α where g_α and h_α are in A and $h_\alpha(u)$ is never zero on U_α. We may assume that $U_\alpha = D(k_\alpha)$ for some k_α in A. Doing the replacement $f(u) = \frac{(k_\alpha g_\alpha)(u)}{(k_\alpha h_\alpha)(u)}$ on $D(k_\alpha \cdot h_\alpha) = D(k_\alpha)$, we may assume that $h_\alpha = k_\alpha$.

Next consider the function $h_\alpha^2 f$. This equals $h_\alpha g_\alpha$ on $D(h_\alpha)$ and both functions are zero on the complement. Therefore $h_\alpha^2 f = h_\alpha g_\alpha$ is in A. By Corollary 1.4.7, $1 = \sum a_\alpha h_\alpha^2$ for some a_α in A. Thus $f = f \cdot 1 = \sum a_\alpha (f h_\alpha^2)$ is in A, which is what we wanted.

It remains to prove $(*)$.

Let X be a space with functions. Then $\star$ defines a bijection Morphism$(X, \text{Spec } A) \to k\text{-Alg-Hom}(A, k[X])$. Let δ_x be evaluation of a function at a point x. Let $f : X \to \text{Spec} A$ be a morphism. Let a be an element of A and x be a point of X. Then $a(f(x)) = (f^*a)(x)$, or, rather, $\delta_{f(x)}$ is the composition $A \xrightarrow{f^*} k[X] \xrightarrow{\delta_x} k$. As the point $f(x)$ is the same as the homomorphism $\delta_{f(x)}$, this shows that f is determined by f^*. Thus $\star$ is injective. Conversely let $\phi : A \to k[X]$ be a k-algebra homomorphism. Define $f \equiv r(\phi)$ by the formula $\delta_{f(x)} = \delta_x \circ \phi$. Using properties (a) and (b) of a space with functions, you do

Exercise 1.5.1. f is a morphism.

Clearly r is an inverse to $\star$. Thus $(*)$ is true and we have constructed all affine varieties up to isomorphism.

Exercise 1.5.2. Show that the category of affine varieties with morphisms is contravariantly equivalent to the category of finitely generated k-algebras with no nilpotents with k-algebra homomorphisms.

In general let X be a subset of a space with functions Y. Then X has an *induced structure of a space with functions*. Explicitly give X the subspace topology. A regular function f on an open set U of X is a function of the following form; there is an open cover $U = \bigcup(X \cap V_\alpha)$ where the V_α are open subsets of Y such that $f(y) = g_\alpha(y)$ on $X \cap V_\alpha$ where g_α is regular on V_α.

Exercise 1.5.3. Check that X is a space with functions and the inclusion $X \hookrightarrow Y$ is a morphism.

With this notation we have

Theorem 1.5.4. *A locally closed subspace of a variety is a variety called a* subvariety. *A closed subspace of an affine variety is affine and in this case a regular function on the subspace lifts to a regular function on the ambient variety.*

Proof. Let X be a locally closed subset of a variety Y. Then X is a closed subset of an open subset Z of Y. Clearly the space with function structures on X induced by Z and Y are the same. Thus we need to know the two cases X open in Y and X closed in Y.

Let $Y = \bigcup Y_i$ where the Y_i are a finite number of open affines. Then $X = \bigcup(X \cap Y_i)$ is a finite open cover of X. As the statements are local on Y we may assume that Y is affine.

Then we want to prove

(a)   if X is closed then X is affine,

(b)   if X is open then X has a finite covering by open affines.

Now let $Y = \operatorname{Spec} A$ where A is a finitely generated k-algebra with no nilpotents. Assume that X is closed. Let I be the ideal of functions in A vanishing identically on X. Clearly set-theoretically $X = \operatorname{Spec}(A/I)$. One simply checks directly from the definitions that

Exercise 1.5.5. $\operatorname{Spec}(A/I)$ has the induced structure as a space with functions. Therefore X is affine if it is closed in an affine.

Assume that X is open. Then $X = \bigcup D(a_i)$ where the a_i are in A. Then the $D(a_i)$ are the affines $\operatorname{Spec}(A_{(a_i)})$ by Exercise 1.3.5. To see that there are only finitely many necessary $D(a_i)$ we use

Lemma 1.5.6. *Any open subset of a variety is quasi-compact.*

This will be proved in the second chapter.

Exercise 1.5.7. Let X be a subspace of a space with functions Y. For any space with functions Z a mapping $Z \to X$ is a morphism if the composition $Z \to X \to Y$ is a morphism.

1.6 $\mathbf{A}^n$ and $\mathbf{P}^n$

By definition $\mathbf{A}^n = \mathrm{Spec}(k[X_1,\ldots,X_n])$. As a k-algebra homomorphism $k[X_1,\ldots,X_n] \to k$ is determined by the images $x_1,\ldots,x_n$ of $X_1,\ldots,X_n$ in k which can be arbitrary, $\mathbf{A}^n = \{(x_1,\ldots,x_n) \in k^n\}$ set-theoretically. As a variety $\mathbf{A}^n$ is called an *affine n-space*.

Furthermore if X is a space with functions and $f : X \to \mathbf{A}^n$ is a mapping given by $f(x) = (f_1(x),\ldots,f_n(x))$ then f is a morphism if and only if each $f_i(x)$ is a regular function on X.

Another fact is that any affine variety X is isomorphic (non-canonically) to a closed subvariety of $\mathbf{A}^n$ corresponding to a surjection $k[X_1,\ldots,X_n] \to A$ where $A = k[X]$. A subvariety of $\mathbf{A}^n$ (or any other affine variety) is called *quasi-affine*.

For instance the punctured affine space $\mathbf{A}^n - \{0\}$.

Lemma 1.6.1.

(a) $k[\mathbf{A}^1 - \{0\}] = k[X_1, X_1^{-1}]$.

(b)   If $n > 1$, $k[\mathbf{A}^n - \{0\}] = k[X_1,\ldots,X_n]$.

Proof. If $n = 1$, $\mathbf{A}^1 - \{0\}$ is the open subvariety $D(X_1)$ of $\mathbf{A}^1$. So $\mathbf{A}^1 - \{0\}$ is the affine $\mathrm{Spec}(k[X]_{(X)} = k[X, X^{-1}])$. Thus $k[X, X^{-1}] = k[\mathbf{A}^1 - \{0\}]$.

If $n \geq 2$, $A^n - \{0\} = D(X_1) \cup \ldots \cup D(X_n)$ and

$$k[D(X_i)] = k[X_1,\ldots,X_n,X_i^{-1}].$$

Thus $k[\mathbf{A}^n - \{0\}] = \bigcap_i k[X_1,\ldots,X_n,X_i^{-1}] = k[X_1,\ldots,X_n]$. $\square$

Exercise 1.6.2. Prove that the quasi-affine variety $\mathbf{A}^2 - \{0\}$ is not affine.

Set-theoretically $\mathbf{P}^n$ is the quotient set

$$(\mathbf{A}^{n+1} - \{0\})/k^* = \{(x_0,\ldots,x_n) \neq 0$$

modulo $(x_0,\ldots,x_n) \sim (\lambda x_0,\ldots,\lambda x_n)$ for any λ in $k - \{0\}\}$. We want to define $\mathbf{P}^n$ as a quotient space with functions. Let $\pi : \mathbf{A}^{n+1} - \{0\} \to \mathbf{P}^n$ send a vector to all vectors with the same direction. A subset U of $\mathbf{P}^n$ is open iff $\pi^{-1}U$ is open. A function f on U is regular iff $\pi^* f$ is regular on $\pi^{-1}U$. We claim that with this definition of a space with functions $\mathbf{P}^n$ is a variety. Let $0 \leq i \leq n$. Let $E_i = \{(x_0,\ldots,x_n)$ in $\mathbf{P}^n \mid x_i \neq 0\}$.

Claim. The E_i form an open covering of $\mathbf{P}^n$ by open affine subvarieties. In fact $E_i = \operatorname{Spec}[k[X_0/X_i, \ldots, X_i/X_i, \ldots, X_n/X_i]]$.

Proof. It suffices to see the second fact. Let $D_i = D(X_i)$ in $\mathbf{A}^{n+1}$. Then $\pi^{-1}E_i = D_i$. So E_i is open and $k[E_i] \approx \{f \in k[X_0, \ldots, X_n, X_i^{-1}]$ such that $f(\lambda x) = f(x)$ for all $\lambda \neq 0\}$. So these f's are homogeneous because k is infinite and hence have the form

$$f \in k[X_0/X_i, \ldots, X_i/X_i, \ldots, X_n/X_i].$$

Thus $B_i \equiv k[X_0/X_i, \ldots, X_n/X_i] = k[E_i]$ and we have a canonical morphism $I : E_i \to \operatorname{Spec} B_i$ which is clearly bijective. We need to check that the inverse of I is a morphism.

Let $\psi : k[X_0, \ldots, X_n, X_i^{-1}] \to B_i$ be the homomorphism sending X_j to X_j/X_i if $j \neq i$ and both X_i and X_i^{-1} to 1. Then this induces a morphism $\operatorname{Spec} B_i \to D_i$. The composition $\operatorname{Spec} B_i \to D_i \xrightarrow{\pi} E_i$ is a morphism which is easily seen to be the inverse of I. Therefore each E_i is affine. Therefore $\mathbf{P}^n$ is a variety called *projective n-space*.

A closed subvariety of a projective space is called *projective*. An arbitrary subvariety of a projective space is called *quasi-projective*. We may compute the global regular functions on $\mathbf{P}^n$. They are just constants.

Lemma 1.6.3. $k[\mathbf{P}^n] = k$.

Proof. If $n = 0$, $\mathbf{P}^0$ is a point and the result is clear. Assume that $n \geq 1$. Then $\pi^\star : k[\mathbf{P}^n] \to k[\mathbf{A}^{n+1} - \{0\}]$ is injective. By Lemma 1.6.1 the last ring is $k[X_0, \ldots, X_{n+1}]$ and $k[\mathbf{P}^n]$ is identified with the homogeneous polynomials of degree zero; i.e. constants. $\qquad\square$

Exercise 1.6.4. Let $f : \mathbf{P}^n \to X$ be a morphism where X is a quasi-affine variety. Prove that the image of f is a single point.

Exercise 1.6.5. Let X be a subset of $\mathbf{P}^n$. Then X is closed if and only if it has the form $\{(X_0, \ldots, X_n) \in \mathbf{P}^n \mid f_i(X_0, \ldots, X_n) = 0$ for some set of homogeneous polynomials $f_i\}$.

1.7 Determinantal varieties

Let X be a variety. Consider a matrix $\alpha = (\alpha_j^i)$ where the coefficients α_j^i are regular functions on X. For each point x of X we can think of $\alpha(x) \equiv (\alpha_j^i(x))$ as the matrix of a linear transformation $\tilde{\alpha}(x) : V \to$

W between two fixed vector spaces. By assumption the matrix $\alpha(x)$ depends regularly on x.

Let r be an integer. Consider the subset $D_r(\alpha) \equiv \{x \in X | \text{rank}\alpha(x) \leq r\}$ of X, where rank denotes the rank of a matrix.

Lemma 1.7.1. $D_r(\alpha)$ *is a closed subvariety of* X.

Proof. We need to explicitly give equations for $D_r(\alpha)$. Let $I_r(\alpha)$ be the set of all determinants of all $(r+1) \times (r+1)$ submatrices of α. Then $I_r(\alpha)$ consists of regular functions on X and $x \in D_r(\alpha) \Leftrightarrow f(x) = 0$ for all f in $I_r(\alpha)$. This shows that $D_r(\alpha)$ is closed. □

If $\alpha = (f_1, \ldots, f_s)$ is just a $1 \times s$ matrix, $D_0(\alpha) = \{f_1(x) = \ldots = f_s(x) = 0\}$.

One way to restate the lemma is

Corollary 1.7.2. *The function* $\text{rank}(\alpha(x))$ *on* X *is lower-semicontinuous.*

2

The preparation lemma
and some consequences

2.1 The lemma

Let f be a non-constant polynomial in variables $X_1, \ldots, X_n$.

Lemma 2.1.1. *After making a linear change of variables of the form* $X_1 \to X_1 + \lambda_1 X_n, \ldots, X_{n-1} \to X_{n-1} + \lambda_{n-1} X_n$, *and* $X_n \to \lambda_n X_n$, $\lambda_n \neq 0$, *where the* λ_i *are constants, we may assume that* f *is monic in* X_n; *i.e.* f *has the form* $X_n^d + \sum\limits_{0 \leq i < d} f_i X_n^i$ *where* $d > 0$ *and the* f_i *are polynomials in the* $n - 1$ *variables* $X_1, \ldots, X_{n-1}$.

Proof. Let d be the total degree of f in all its variables. Write $f = f_d + g$ where f_d is homogeneous of degree d and the total degree of g is $< d$. Consider $f' = f(X_1 + \lambda_1 X_n, \ldots, X_{n-1} + \lambda_{n-1} X_n, \lambda_n X_n)$ for variables λ_i. We have $f' = (\text{terms in } X_n \text{ of degree} < d) + f_d(\lambda_1, \ldots, \lambda_n) X_n^d$. Thus we want to choose $\lambda_1, \ldots, \lambda_n$ such that $f_d(\lambda_1, \ldots, \lambda_n) = 1$ where we know that f_d is a non-zero homogeneous polynomial. By homogeneity it is enough to find $\mu_1, \ldots, \mu_n$ such that $a \equiv f_d(\mu_1, \ldots, \mu_n) \neq 0$ because it may take $\lambda = (\frac{1}{a})^{1/d} \cdot \mu$.

Thus we want to show that a non-zero polynomial g in $X_1, \ldots, X_n$ defines a non-zero function on $\mathbb{A}^n$ (which is a special case of the nullstellensatz which has not been proven yet). The proof is by induction on n. If $n = 1$, the number of zeroes of $g(X_1)$ is finite but k is infinite.

Thus this case is clear. If $n > 1$, write $g = \sum\limits_{0 \leq i \leq d} g_i X_n^i$ where the g_i are polynomials in $X_1, \ldots, X_{n-1}$ and $g_d \not\equiv 0$. By induction we can find $\mu_1, \ldots, \mu_{n-1}$ such that $g_d(\mu_1, \ldots, \mu_{n-1}) \neq 0$. Thus $g(\mu_1, \ldots, \mu_{n-1}, X_n)$ is a non-zero polynomial in X_n. By the case $n = 1$ we may find μ_n such that $g(\mu_1, \ldots, \mu_n) \neq 0$. □

Next we will give the proof of Lemma 1.4.1; i.e. If A is a non-zero finitely generated k-algebra then we have an injection $B \subset A$ such that B is a polynomial ring and A is a B-module of finite type.

Proof. Let $k[X_1, \ldots, X_n] \to A$ be a surjective k-algebra homomorphism with kernel I. If $I = 0$ then take $B = A$. Otherwise let i be a non-zero element of I. By assumption i is not constant. By Lemma 2.1.1, we may assume that i is monic of degree $d > 0$ in X_n after changing coordinates. Let $A' =$ image of $k[X_1, \ldots, X_{n-1}]$ in A.

Claim. A is an A'-module of finite type.

Note that we have a surjection $k[X_1, \ldots, X_n]/(i) \to A$ and the first ring is a free $k[X_1, \ldots, X_{n-1}]$-module with basis $1, \ldots, X_n^{d-1}$ modulo(i) by long division. The claim follows immediately.

As A' has fewer generators than A, by induction on n (the case $n = 0$ is trivial) we may find a polynomial ring $B \subset A'$ such that A' is a B-module of finite type. Hence by the claim A is a B-module of finite type. □

We will need to know a special case of this argument.

Lemma 2.1.2. *Let i be a non-constant regular function on $\mathbf{A}^n$. We may find a morphism $\pi : \{i = 0\} \to \mathbf{A}^{n-1}$ which is surjective such that, via π^*, $k[\{i = 0\}]$ is a $k[\mathbf{A}^{n-1}]$-module of finite type.*

Proof. As above $B = k[X_1, \ldots, X_n]/(i)$ is a finite $k[\mathbf{A}^{n-1}]$-module where $X_1, \ldots, X_{n-1}$ are the coordinate functions on $\mathbf{A}^{n-1}$. Thus $k[\{i = 0\}] = B/\sqrt{0}$ is a finite type module over $k[\mathbf{A}^{n-1}]$. Now if $A = k[\mathbf{A}^{n-1}]$ we have the inclusion $A \subset B$, then $\{i = 0\} = \operatorname{Spec} B/\sqrt{0} = k\text{-Hom}(B/\sqrt{0}, k) \approx k\text{-Alg-Hom}(B, k) \to k\text{-Alg-Hom}(A, k) = \mathbf{A}^{n-1}$ is surjective by Lemma 1.4.2. □

Thus we have finished the proof of the nullstellensatz.

2.2 The Hilbert basis theorem

A ring A is *noetherian* if any ideal of A is finitely generated. This condition is easily seen to be equivalent to the following by an induction on the number of generators; any submodule of a finitely generated A-module M is finitely generated.

Exercise 2.2.1. Prove this equivalence.

Hilbert's basis theorem is

Theorem 2.2.2   *A finitely generated k-algebra A is noetherian.*

Proof. Let $\psi : k[X_1, \ldots, X_n] \to A$ be a surjection. Let I be an ideal of A. To show that I is finitely generated it suffices to show the same for the ideal $I' = \psi^{-1}(I)$ in the polynomial ring. If $I' = 0$ there is no problem. Otherwise there exists a non-zero polynomial f in I'. By Lemma 2.1.1 we may assume that f is monic in X_n.

By induction we may assume that $k[X_1, \ldots, X_{n-1}] = B$ is known to be noetherian. Consider $I'/(f) \subset k[X_1, \ldots, X_n]/(f)$. This is a B-submodule of a B-module which is finitely generated. Then $I'/(f) = (\bar{g}_1 B + \ldots + \bar{g}_r B)$ where $g_1, \ldots, g_r$ are elements of I'. Clearly $I = (f, g_1, \ldots, g_r)$ and we are finished. $\qquad\Box$

We may now verify Lemma 1.5.6. By the previous reasoning we need to show that an open subset U of an affine variety $\operatorname{Spec} A$ is quasi-compact. Let $U = \bigcup D(f_i)$ for f_i in A. Then the complement of U is $\operatorname{zeroes}((f_i)_{i \in I}) = \operatorname{zeroes}((f_{i_1}, \ldots, f_{i_r}))$ for some finite number r by Theorem 2.2.2. Thus $U = D(f_{i_1}) \cup \ldots \cup D(f_{i_r})$ is a finite subcovering. This is what we want to show.

A topological space such that every open subset is quasi-compact is called *noetherian*. As a variety is the finite union of open affines,

Theorem 2.2.3.   *Any variety has a noetherian topology.*

There is a chain condition that characterizes noetherian space.

Lemma 2.2.4   *Let X be a topological space. Then X is noetherian if and only if any descending chain $X_1 \supseteq X_2 \supseteq \ldots$ of closed subsets becomes stationary eventually: $X_r = X_{r+1} = \ldots$ for some r.*

Proof. Assume that X is noetherian and we have such a chain. The open subset $U = X - \bigcap_i X_i$ is quasi-compact. Now U has the chain $X_1 \cap U \supseteq \ldots$ of closed subsets which has empty intersection. Thus $X_r \cap U = \emptyset$ for some r. This means $X_r \subseteq \bigcap_i X_i$ and thus $X_r = X_{r+1} \ldots$.

Conversely if $U_1 \supseteq U_2 \supseteq$ is a descending chain of closed subsets of U such that $\bigcap_i U_i = \emptyset$, consider $X_i =$ closure of U_i in X. Then $X_r = X_{r+1} = \ldots$ for some r. As $U_i = X_i \cap U$ we have $U_r = U_{r+1} = \ldots = \emptyset$. Thus U is quasi-compact. $\square$

In the next section we will see that noetherian spaces have special-properties.

2.3 Irreducible components

A topological space X is *irreducible* if whenever $X = X_1 \cup X_2$ where X_1 and X_2 are closed subspaces then $X = X_1$ or X_2. An irreducible topological space is necessarily connected but not conversely as the following example shows. The plane curve $x(y - x^2) = 0$ is connected but not irreducible. This curve is the union of the line $x = 0$ and the parabola $y = x^2$ which are irreducible but they meet at the origin $\{0, 0\}$.

We have an algebraic criterion for an affine variety to be irreducible.

Lemma 2.3.1. *An affine variety X is irreducible iff the ring $k[X]$ is an integral domain.*

Proof. Let $X_i = \text{zeroes}(I_i)$ for I_i in $k[X]$ be two closed subsets of X. Assume that $X = X_1 \cup X_2 = \text{zeroes } (I_1 \cdot I_2)$; or, equivalently, $I_1 \cdot I_2 = 0$. Thus if $k[X]$ is integral, $I_1 = 0$ or $I_2 = 0$, or, equivalently $X_1 = X$ or $X_2 = X$. Thus X is irreducible if $k[X]$ is integral and the converse is just as obvious. $\square$

Before we discuss components we will do two lemmas.

Lemma 2.3.2. *Let Z be an irreducible closed subspace of a topological space X. If Z is contained in the finite union $W_1 \cup \ldots \cup W_p$ for some closed subsets of X, then $Z \subset W_i$ for some i.*

Proof. Z equals the finite union $(Z \cap W_1) \cup \ldots \cup (Z \cap W_p)$ of closed subsets. As Z is irreducible, $Z = Z \cap W_i$ for some i. Thus, $Z \subset W_i$. $\square$

We will next see that a noetherian topological space is not too far removed from being irreducible.

Lemma 2.3.3. *Any noetherian topological space X is a finite union of irreducible closed subspaces.*

Proof. If X is irreducible, there is nothing to prove. Otherwise, X is the union $Z \cup W$ of two proper closed subsets. If Z and W are finite unions of irreducible closed subsets, then so is X.

Therefore, if X were not a finite union of irreducible closed subsets, then we could find a proper closed subset X_1 (either Z or W) with the same property. Continuing in this way, one could produce an infinite strictly decreasing chain $X \supsetneq X_1 \supsetneq X_2 \supsetneq \ldots$ of closed subsets, which we know is impossible as X is noetherian. Thus, X must be a finite union of irreducibles. □

To be more precise about the structure of a noetherian space, one introduces the idea of a component. A *component* of X is just a maximal irreducible closed subspace of X. In our case, this is really not very abstract. In fact, we have

Proposition 2.3.4. *A noetherian space X has only finitely many components. The union of these components is all of X. Furthermore, if $X = X_1 \cup \ldots \cup X_n$ is a finite union of closed irreducible subspaces, then the components of X are exactly the X_i's which are not contained in any other.*

Proof. If we prove the last statement, it will be enough because Lemma 2.3.3 says that any noetherian space is such a union. Let Z be any irreducible closed subset of X. By Lemma 2.3.2, $Z \subset X_i$ for some i. Therefore, any component (maximal Z) must be one of the X_i, which is not contained in another. Conversely, the big X_i's must be maximal irreducibles; i.e. they are components. □

Next, we will study the algebraic interpretation of the components.

Proposition 2.3.5. *Let X be an affine variety.*

(a) *A closed subset Z of X is a component of X if and only if the ideal $\{f \in k[X] | f|_Z = 0\}$ is a minimal prime ideal of $k[X]$.*

(b) *There is a one-to-one correspondence between the components of X and the minimal prime ideals of $k[X]$.*

(c)   *If $X_1, \ldots, X_r$ are the components of X, the zero ideal is the intersection $\bigcap$ ideal(X_i) of finitely many prime ideals.*

Proof. We first recall that a component Z of X is a maximal irreducible closed subset. By Lemma 2.3.1, a closed subset Z of X is irreducible $\Longleftrightarrow k[x]/$ ideal(Z) is an integral domain $\Longleftrightarrow$ ideal(Z) is a prime ideal. The rest follows formally.   □

Exercise 2.3.6. Show that a topological space X is irreducible if and only if any non-empty open subset of X is dense.

Exercise 2.3.7. Let X be an irreducible topological space. Show that any open subset of X must be irreducible.

Exercise 2.3.8. Let X be a topological space which has an open covering by irreducible subspaces. Prove that X is connected if and only if it is irreducible.

Exercise 2.3.9. Let $X = \bigcup U_i$ be an open cover of a noetherian topological space X. Show that the components of X are precisely the closure of the components of the U_i.

2.4 Affine and finite morphisms

A morphism $f : X \to Y$ is *affine* if there is an open cover $Y = \bigcup U_i$ by affines U_i such that the open subsets $f^{-1}(U_i)$ of X are affine. The morphism f is *finite* if we can furthermore choose the cover such that each $k[f^{-1}(U_i)]$ is a $k[U_i]$-module of finite type.

If Y is a point $P, f : X \to P$ is affine if and only if X is affine and $f : X \to P$ is finite if and only if X is a finite set. As for embedding we have

Lemma 2.4.1.

(a)   *Let $i : X \hookrightarrow Y$ be the inclusion of a closed subvariety. Then i is finite.*

(b)   *Let f be a regular function on a variety X. Then the inclusion j. $D(f) \hookrightarrow X$ is affine.*

Exercise 2.4.2. Prove Lemma 2.4.1.

Topologically finite morphisms are special.

Lemma 2.4.3. *Let $f : X \to Y$ be a finite morphism. Let Z be a closed subset of X. Then its image $f(Z)$ is closed.*

Proof. As the statement is local on Y we may assume that X and Y are affine and $k[X]$ is a $k[Y]$-module of finite type. Consider the surjection $i^* : k[X] \to k[Z]$. Let I be the kernel of the composition $\psi : k[Y] \xrightarrow{f^*} k[X] \to k[Z]$. Then we have an inclusion $A = k[Y]/I \subset k[Z] = B$ and B is an A-module of finite type.

By definition a point of Y is contained in the image $f(Z)$ if and only if $\delta_y : k[Y] \to k$ is the composition of ψ with $\delta_z : k[Z] \to k$ for some z in Z. Thus $f(Z) \subset \text{zeroes}(I)$ and it is enough to show the reverse inclusion. In other words if y is in $\text{zeroes}(I)$ then δ_y gives a homomorphism $A \to k$ which by Lemma 1.4.2 we can extend to a homomorphism $\delta_z : B \to k$ where z is a point of Z. Thus $f(z) = y$ and we are done. □

We have another property of finite morphisms

Lemma 2.4.4. *Let $f : X \to Y$ be a finite morphism and let $Z_1 \subsetneq Z_2$ be two closed subsets of X where Z_2 is irreducible. Then $f(Z_1) \subsetneq f(Z_2)$.*

Before we prove this let us see what it says in a special case.

Corollary 2.4.5. *If $f : X \to Y$ is a finite morphism and y is a point of Y then the fiber $f^{-1}(y)$ is finite.*

Proof. It will suffice to show that each irreducible component of Z_2 of $f^{-1}(y)$ is a point. Let z be a point of Z_2. Let $Z_1 = \{z\}$. Then $f(Z_1) = f(Z_2) = \{y\}$. Hence $Z_1 = Z_2$. □

Proof of Lemma 2.4.4. It clearly suffices to do the case where X and Y are affine and $k[X]$ is a $k[Y]$-module of finite type. Furthermore, we may assume that $Z_2 = X$ and $Y = f(X)$. Thus $k[X]$ is an integral domain and is finitely generated as a module over the subring $k[Y]$. We need to show that there is a non-constant regular function g on Y such that f^*g vanishes on Z_1 because this will show that $f(Z_1) \subseteq \text{zeroes}(g) \subsetneq f(X) = Y$. As $Z_1 \subsetneq X$ there is a non-constant regular function h on X which vanishes on Z_1. Consider the $k[Y]$-submodule of $k[X]$ generated by $(h^i)_{i \in \mathbf{N}}$. Then as this is finitely generated over $k[Y]$, for $i \gg 0$, $h^i = \sum_{0 \le j < i} (f^*g_j)h^j$ where the g_j are in $k[Y]$. Factoring out h's we may assume that $g_0 \neq 0$. Now $h | f^*g_0 = $ regular so f^*g_0 vanishes on Z_1. Thus $g_0 = g$ works. □

2.5 Dimension

A topological space X is said to have dimension $\leq n$ if, for any strictly decreasing chain $Z_p \subsetneq Z_{p-1} \subsetneq \ldots \subsetneq Z_0$ of non-empty irreducible closed subsets of X, the length p of the chain $\leq n$. The minimum n that satisfies this condition is called the *dimension* of X. Clearly, for a noetherian topological space X, $\dim X = $ maximum of dimensions of the components of X.

In algebraic geometry, this naive notion of dimension is quite good. An affine variety of dimension zero is a finite collection of points. An affine variety of dimension one is a finite union of irreducible curves plus a finite number of isolated points.

We will first study how finite morphisms behave with respect to dimension.

Lemma 2.5.1. *Let $f : X \to Y$ be a finite surjective morphism between varieties. Then $\dim X = \dim Y$.*

Proof. Let $Z_p \supsetneq \ldots \supsetneq Z_0$ be a chain of non-empty irreducible closed subsets of X. Set $W_i = f(Z_i)$. Then $W_p \supseteq \ldots \supseteq W_0$ is a chain of non-empty irreducible closed (Lemma 2.4.3) subsets of Y. By Lemma 2.4.4 we have $W_q \neq W_{q'}$, if $q \neq q'$. Thus $\{W_q\}$ is strictly decreasing. Therefore $\dim X \leq \dim Y$.

Conversely let $W_p \supsetneq \ldots \supsetneq W_0$ be a sequence in Y. We need to see that we can find a sequence of Z_i in X such that $f(Z_i) = W_i$. The variety $f^{-1} W_p = K_1 \cup \ldots \cup K_i$ where the K_i's are the irreducible components of $f^{-1} W_p$. Thus as f is surjective $W_p = f(K_1) \cup \ldots \cup f(K_i)$ where the $f(K_j)$ are closed and irreducible. As W_p is irreducible, $W_p = f(K_j)$ for some j. Set $Z_p = K_j$. Consider the finite surjective morphism $f : Z_p \to W_p$. By induction on p we may find the required $Z_{p-1} \supset \ldots \supset Z_0$ in Z_p. This constructs the required $\{Z_j\}$. $\square$

Next we will check that we have the right definition of dimension in the simplest case.

Theorem 2.5.2. $\dim \mathbf{A}^n = n$.

Proof. Let $\mathbf{A}^n = \{(x_1, \ldots, x_n)\}$. Set $Z_i = \{x \in \mathbf{A}^n | x_{i+1} = \ldots = x_n = 0\}$. Then $\mathbf{A}^n = Z_n \supsetneq \ldots \supsetneq Z_0$ is a strictly descending chain of closed subsets. As $Z_p \approx \mathbf{A}^p$ each Z_p is irreducible. Therefore $\dim \mathbf{A}^n \geq n$ and we need to prove that if $Z_p \supsetneq \ldots \supsetneq Z_0$ is a strictly descreasing chain of closed irreducible subsets of $\mathbf{A}^n$ then $p \leq n$. As $Z_{p-1} \subsetneq \mathbf{A}^n$ is closed we may find a non-constant regular function g such that $Z_{p-1} \subset$ zeroes(g). If we show that dim zeroes$(g) \leq n - 1$, then $p - 1 \leq n - 1$ and we will be done.

By Lemma 2.1.2, we have a finite surjective morphism zeroes$(g) \rightarrow \mathbf{A}^{n-1}$. By Lemma 2.5.1 dim zeroes$(g) = \dim \mathbf{A}^{n-1} = n - 1$ by induction.

$\square$

From the proof we have

Corollary 2.5.3. *Let $g \neq 0$ be a non-constant regular function on $\mathbf{A}^n$, then* $\dim(g = 0) = n - 1$.

We also get

Corollary 2.5.4. *Any variety has finite dimension.*

Proof. Let X be a variety. If X is affine then it is isomorphic to a closed subset of some $\mathbf{A}^n$. Thus $\dim X \leq \dim \mathbf{A}^n = n$ is finite. To reduce to the affine case we use an open affine cover $X = X_1 \cup \ldots \cup X_r$. Then we finish the proof with

Lemma 2.5.5. $\dim X = \max \dim X_i$.

Exercise 2.5.6. Prove this lemma.

Exercise 2.5.7. Let $Z \subset X$ be a closed subset of an irreducible variety X. Then $Z = X$ iff $\dim Z = \dim X$.

2.6 Hypersurfaces and the principal ideal theorem

A *hypersurface* in $\mathbf{A}^n$ is a closed subset of the form $g = 0$ where g is a non-constant regular function on $\mathbf{A}^n$. These affine varieties are the first ones to be studied in detail. We assume that the reader knows that the polynomial ring $= k[\mathbf{A}^n]$ is a unique factorization domain and the units in $k[\mathbf{A}^n]$ are just the non-zero constants.

Theorem 2.6.1. *Let $g = $ constant $\prod_{1 \leq i \leq d} g_i^{n_i}$ where the g_i are inequivalent irreducible polynomials.*

(a) *The components of the hypersurface $g = 0$ are the hypersurfaces $g_i = 0$.*

(b) $k[g = 0] = k[\mathbf{A}^n]/(\prod_{1 \leq i \leq d} g_i)$.

(c) *Each component of a hypersurface has dimension $n-1$. Conversely if X is a closed subvariety of $\mathbf{A}^n$ whose components all have dimension $n - 1$ then X is a hypersurface.*

Proof. Clearly $(g = 0) = (g_1 = 0) \cup \ldots \cup (g_d = 0)$. To show (a) we need to see that (1) the hypersurfaces $g_i = 0$ are irreducible and (2) they do not contain one another. As g_i is irreducible, the ideal (g_i) is prime. Thus $A = k[\mathbf{A}^n]/(g_i)$ is an integral domain.

Hence the set $(g_i = 0) = \operatorname{Spec} A$ is irreducible and shows that $k[g_i = 0)] = k[\mathbf{A}^n]/(g_i)$. This proves (1). For (2) if $(g_i = 0) \subset (g_j = 0)$ then g_j vanishes on $g_i = 0$. Thus $g_i | g_j$ which is impossible. Thus (a) is true.

For (b) we need to see that the polynomial functions h on $\mathbf{A}^n$ that vanish on $g = 0$ are exactly those divisible by $g_1 \ldots g_d$. If h is divisible by $g_1 \ldots g_d$, then h is zero on $\bigcup(g_i = 0) = (g = 0)$. Conversely if h vanishes on $(g = 0)$, then h vanishes on each $g_i = 0$. By the first paragraph this means that each g_i divides h. As $k[\mathbf{A}^n]$ is a UFD then this means that $g_1 \ldots g_d$ divides h.

For (c) the first statement follows from Corollary 2.5.3. Conversely we need only treat the case where $X \subset \mathbf{A}^n$ is an irreducible closed subvariety of dimension $n - 1$. As $X \neq \mathbf{A}^n$ for dimension reasons we have a non-constant regular function g on $\mathbf{A}^n$ such that $X \subset (g = 0)$ but $\dim X = \dim(g = 0)$. Thus X is a maximal irreducible closed subset of $g = 0$. Thus X is a component of $(g = 0)$. Thus X is a hypersurface by (a). $\square$

One consequence of the theorem is a general fact about varieties.

Corollary 2.6.2. *If g is a non-zero regular function on $\mathbf{A}^n$ such that $\dim(g = 0) \leq \dim(\mathbf{A}^n) - 2$ then g is a unit in $k[\mathbf{A}^n]$.*

We next can prove the general result

Theorem (principal ideal theorem) 2.6.3. *Let g be a non-zero regular function on an irreducible variety X. Then each component of the closed subset $(g = 0) = \{x \in X | g(x) = 0\}$ has dimension $\dim(X) - 1$.*

Proof. We begin with

Lemma 2.6.4. *If X is any irreducible affine variety, Corollary 2.6.2 is true when X replaces $\mathbf{A}^n$.*

Proof. The idea is to reduce to the case where $X = \mathbf{A}^n$ by the preparation lemma. Let $f : X \to \mathbf{A}^n$ be a finite surjective morphism such that $f^* : k[\mathbf{A}^n] \subset k[X]$ makes $k[X]$ a finite $k[\mathbf{A}^n]$-module. By the proof of Lemma 2.4.4, g satisfies a non-trivial polynomial relation $\sum a_i g^i = 0$ where a_i are in $k[\mathbf{A}^n]$. Let $H(X_{n+1}) = \sum\limits_{0 \leq i \leq d} a_i X_{n+1}^i$ where $a_0 \neq 0$ is an irreducible polynomial in $k[\mathbf{A}^n][X_{n+1}] = k[\mathbf{A}^{n+1}]$ such that $H(g) = 0$. Let $h : X \to \mathbf{A}^{n+1}$ be the morphism given by $(f(x), g(x))$. As f is finite h is finite. Then $h(X) \subseteq (H = 0)$ and $h(X)$ is a closed subset of dimension $\dim X = n$ which equals $\dim H = (n + 1) - 1$. As $(H = 0)$ is irreducible, $h : X \to (H = 0)$ is a surjective finite morphism. Now $(g = 0) = h^{-1}((H = 0) \cap (X_{n+1} = 0)) = h^{-1}((a_0 = 0) \times \{0\})$. Thus $\dim(g = 0) = \dim((a_0 = 0)) \leq n - 2$ by hypothesis. By Corollary 2.6.2 a_0 is a unit on $\mathbf{A}^n$. Now $a_0 = (- \sum\limits_{1 \leq i \leq d} a_i g^{i-1}) \cdot g$. Thus g is never zero because a_0 is never zero. Thus $\frac{1}{g}$ is regular. $\qquad\square$

For Theorem 2.6.3, assume that Z is a component of $g = 0$ where $\dim Z \leq \dim(X) - 2$. We can find an open affine variety U of X such that $U \cap \{g = 0\} = Z \cap U$ is non-empty.

Let h be a regular function on X such that h is not identically zero on $Z \cap U$ and zero on the other components of $(g = 0) \cap U$. Then if we take U to be $D(h)$ then $Z \cap U$ is the only component of $g|_U = 0$. Now assume that we prove

Lemma 2.6.5. *If U is a non-empty open subset of an irreducible variety X, then $\dim U = \dim X$.*

Then we will have $\dim U \cap Z = \dim Z \leq \dim X - 2 = \dim U - 2$. Hence by Lemma 2.6.4, $g|_U$ is a unit. Thus $U \cap Z = (g = 0) \cap U$ is empty, which gives a contradiction which proves the theorem. Therefore we need to prove the lemma.

For the lemma clearly $\dim U \leq \dim X$ as we may take the closure of a chain in U. If $X = \mathbf{A}^n$ then $\dim U = \dim X$ because we may take a chain in U of the form $\{(\text{point of } U) \subset \text{line} \subset \text{plane} \subset \ldots \mathbf{A}^n\} \cap U$. If X is affine we have a finite surjective morphism $f : X \to \mathbf{A}^n$ by the lemma. By Lemma 2.4.4 $f(X - U)$ is a proper closed subset of $\mathbf{A}^n$. Let V be its complement. Let $U' = f^{-1}(V)$. Then $f : U' \to V$ is a finite surjective morphism. Thus $\dim U' \leq \dim U \leq \dim X$ and $\dim U' = \dim V = \dim \mathbf{A}^n = \dim X$. Thus $\dim U = \dim X$. We can reduce the general to the above. $\qquad\square$

Exercise 2.6.6. Let $H = (XY - Z^2 = 0)$ in $\mathbb{A}^3$. Prove that the line $L = (X = Z = 0)$ is contained in H and $\dim L = \dim(H) - 1$ but the ideal of regular functions on H vanishing on L is not principal.

As a final test for our topological dimension theory we will check

Theorem 2.6.7. *Let $Z_p \supsetneq \ldots \supsetneq Z_0$ be a strictly decreasing chain of non-empty closed irreducible subset of a variety. If $\{Z_i\}$ cannot be refined by another such chain then $\dim Z_i = i$ for all i.*

Proof. Clearly we may assume that Z_0 is a point p. Let U be an open affine neighborhood of p. $\{Z_i \cap U\}$ is such a chain in U and $\dim(Z_i) = \dim(Z_i \cap U)$. Thus we may assume that X is affine. Assume that $\dim Z_i = i$, we want to show that $\dim Z_{i+1} = i + 1$. Now $\dim Z_{i+1} > \dim Z_i = i$. Assume that $\dim Z_{i+1} \geq i + p$ for some $p \geq 2$. Let g be a non-zero regular function on Z_{i+1} which vanishes on Z_i. Then $Z_i \subset (g = 0) \subsetneq Z_{i+1}$. Let Z' be a component of $(g = 0) \supset Z_i$. Then $\dim Z' = i + (p - 1) > \dim Z_i = i$ so $Z_i \subsetneq Z'$ and the chain $Z_p \supsetneq \ldots \supsetneq Z_{i+1} \supsetneq Z' \supsetneq Z_i \supsetneq \ldots \supsetneq Z_0$ is a refinement. Therefore $\dim Z_{i+1} = i + 1$ if $\{Z_i\}$ cannot be refined. $\qquad\square$

A useful little fact is the following.

Lemma 2.6.8. *Let X be a closed subset of an affine variety Y. Assume that each component of X (respectively Y) has the same dimension. Then there are regular functions $f_1, \ldots, f_r$ on Y such that X is a union of components of $zeroes(f_1, \ldots, f_r)$ and each component of $zeroes(f_1, \ldots, f_r)$ has dimension $\dim Y - r = \dim X$.*

Proof. If $\dim X < \dim Y$ we need to find a function f_1 on Y which vanishes on X and does not vanish at $\{y_1, \ldots, y_d\}$ which is a finite set of points meeting each component of $Y - X$. If we do this we will have by the principal ideal theorem that $Y' = zeroes(f_1)$ has all its components of dimension $\dim Y - 1$. Then by induction we have $g_1, \ldots, g_{r-1}$ on Y' such that the conclusion holds. Now just lift $g_1, \ldots, g_{r-1}$ to regular functions $f_2, \ldots, f_r$ on Y and we will be done.

To find f_1 for each i let h_i be a regular function vanishing on the closed subset $X \cup \{y_1, \ldots, y_i, \ldots, y_d\}$ with value one at y_i. Now take $f_1 = \sum_i h_i$. $\qquad\square$

3

Products;
separated and complete varieties

3.1 Products

Let X and Y be two spaces with functions. We have a categorical product of X and Y.

Lemma 3.1.1. *There exists a space with functions $X \times Y$ together with two morphisms $\pi_X : X \times Y \to X$ and $\pi_Y : X \times Y \to Y$ such that for any space with functions Z the mapping* Morphism$(Z, X \times Y) \to$ Morphism$(Z, X) \times$ Morphism(Z, Y) *which sends f to $(\pi \circ f, \pi_Y \circ f)$ is bijective.*

Proof. As a set $X \times Y$ is just the set-theoretic product but $X \times Y$ does not have the product topology. In fact its topology has more open subsets than the product topology. As a basis of open subsets, we have the form $O = \{(u, v) \in U \times V | f(u, v) \neq 0\}$ where U is open in X and V is open in Y and $f(u, v) = \sum_{\text{finite}} g_i(u) h_i(v)$ where the g_i are regular functions on U and the h_i are regular functions on V. Clearly we must have all of these open subsets if π_X and π_Y are morphisms and $X \times Y$ is a space with functions. Also if f' is another such function then $F(u, v) = \frac{f'(u,v)}{f(u,v)}$ must be regular on any open subset O' of O. A regular function G on any open subset P of $X \times Y$ is a k-valued function such that $P = \bigcup O_{i'}$ where $G|_{O_i'}$ has the form F as above.

Exercise 3.1.2. Prove in detail that this space with functions $X \times Y$ satisfies the property of the lemma. $\qquad\qquad\qquad\qquad\qquad\qquad$ □

If U and V are subspaces with functions of X and Y, then $U \times V$ is a subspace with functions of $X \times Y$. This may be checked easily from the universal mapping property of products and subspaces. Here one should note that $U \times V$ is open in $X \times Y$ if U and V are open and similarly with closed.

Returning to algebraic geometry we have

Theorem 3.1.3.

(a) *If X and Y are varieties then $X \times Y$ is a variety.*

(b) *If X and Y are affine, then $X \times Y$ is affine and $k[X \times Y] = k[X] \otimes_k k[Y]$.*

Proof. If $X = X_1 \cup \ldots \cup X_r$ and $Y = Y_1 \cup \ldots \cup Y_s$ are affine open covers then $X \times Y = \bigcup_{i,j} X_i \times Y_j$ is a finite open covering. If (b) is true, each $X_i \times Y_j$ is affine and (a) follows.

To prove (b) we will use a trick to avoid the seemingly difficult description of the product above.

Step 1. The ring $k[X] \otimes_k k[Y]$ is a finitely generated k-algebra with no nilpotents.

Proof. Clearly the tensor products are generated by $X_i \otimes 1$ and $1 \otimes Y_i$ where $\{X_i\}(\{Y_i\})$ are generators of $k[X](k[Y])$. Then the tensor product is finitely generated because its factors are. To show that the tensor product has no nilpotents it suffices to show that it injects into the ring of functions $X \times Y$ by the homomorphism and sends $\sum f_i \otimes g_i$ to $\sum f_i(x)g_i(y)$. Thus we need to check that if $f_1, \ldots, f_i$ are linearly independent functions on X and $g_1, \ldots, g_j$ are linearly independent functions on Y then $(f_p(x)g_q(y))$ are linearly independent functions on the product $X \times Y$.

To do this assume that we have a linear relation $\sum_{p,q} \mu_{p,q} f_p(x)g_q(y) = 0$ where the $\mu_{p,q}$'s are constants. Then when x is fixed we have a relation $\sum_q (\sum_p \mu_{p,q} f_p(x)) g_q(y) = 0$. So $\sum_p \mu_{p,q} f(x) = 0$ for x arbitrary as the g_q are linearly independent. By the linear independence of the f_p we have $\mu_{p,q} = 0$ for all p, q. This finishes Step 1.

Step 2. Use the universal mapping property of the tensor product.

Let A and B be two k-algebras. The tensor product $A \otimes_k B$ is a k-algebra with two homomorphism $\psi_A : A \to A \otimes_k B$ and $\psi_B : B \to A \otimes_k B$ given by $\psi_A(a) = a \otimes 1$ and $\psi_B(b) = 1 \otimes b$. Composing with the ψ's gives a bijection k-Alg-Hom$(A \otimes_k B, C) \overset{\approx}{\to} k$-Alg-Hom$(A, C) \times k$-alg-hom (B, C) for any k-algebra C.

Geometrically we have two morphisms $Z \equiv \text{Spec}(k[X] \otimes k[Y]) \overset{\pi_X}{\to} X$ (resp. $\overset{\pi_Y}{\to} Y$). Let W be a space with functions. Then

$$\text{Morphism}(W, Z) \cong k\text{-Alg-Hom}(k[X] \otimes_k k[Y], k[W])$$

$$|\wr$$

$$k\text{-Alg-Hom}(k[X], k[W]) \times k\text{-Alg-Hom}(k[Y], k[W])$$

$$|\wr$$

$$\text{Morphism}(W, X) \times \text{Morphism}(W, Y)$$

is the bijection given by $f \to (\pi_X \circ f, \pi_Y \circ f)$. Thus Z is an affine variety with $k[Z] = k[X] \otimes_k k[Y]$ which has the required mapping property of $X \times Y$. Thus (b) is true. $\qquad\square$

The moral is: one computes easily with tensor products of k-algebras to prove the relevant properties about products, but set-theoretic properties about products are easy geometrically.

Exercise 3.1.4. Prove that the products of quasi-affine varieties are quasi-affine.

Exercise 3.1.5. Show $\mathbf{A}^n = \mathbf{A}^1 \times \ldots \times \mathbf{A}^1$ (n times).

Exercise 3.1.6. If X and Y are varieties then $\dim(X \times Y) = \dim X + \dim Y$ and $X \times Y$ is irreducible if X and Y are. (Hint: for the first part reduce to the case of two affine spaces).

3.2 Products of projective varieties

The main result of this section is

Theorem 3.2.1. (Segre.) *The product $\mathbf{P}^n \times \mathbf{P}^m$ of two projective spaces is isomorphic to a closed subvariety of $\mathbf{P}^{nm+n+m}$.*

Proof. $\mathbf{P}^n$ is set-theoretically the set of lines through 0 in the vector space k^{n+1}. The embedding sends $(\ell_1 \subset k^{n+1}, \ell_2 \subset k^{m+1})$ to $\ell_1 \otimes_k \ell_2$ in $k^{n+1} \otimes k^{m+1} = k^{mn+m+n+1}$. The image is {lines consisting of tensors of rank one}. This is what the embedding is in terms of modern linear algebra but we will need to write it in coordinates.

Let $(x_0, \ldots, x_n)$ and $(y_0, \ldots, y_m)$ be homogeneous coordinates of a point in $\mathbf{P}^n$ and $\mathbf{P}^m$. Consider $F(x, y) = (z)$ where $Z_{i,j} = X_i \cdot Y_j$. We will use $(Z_{i,j})_{\substack{0 \leq i \leq n \\ 0 \leq j \leq m}}$ as homogeneous coordinates in $\mathbf{P}^{nm+n+m}$. Now $F(\lambda x, \mu y) = \lambda \mu F(x, y)$ and $Z_{i,j} \neq 0$ if $X_i \neq 0$ and $Y_j \neq 0$. So F defines a mapping $S : \mathbf{P}^n \times \mathbf{P}^m \to \mathbf{P}^{nm+n+m}$. It is obvious that S is a morphism as $D(X_i) \times D(Y_j) = \{(x_0, \ldots, x_{i-1}, 1, x_{i+1}, \ldots, x_n)\} \times \{(y_0, \ldots, y_{j-1}, 1, y_{j+1}, \ldots, y_m)\} \xrightarrow{S_{i,j}} D(Z_{ij}) = \{(z_{ij} = 1), \text{ other coordinates arbitrary}\}$ is regular as it is given by a polynomial in the local coordinates.

The image of S is the closed subset T of $\mathbf{P}^{n \times m+n+m}$ given by rank $(Z_{i,j}) \leq 1$; i.e. $\det \begin{pmatrix} Z_{i_1,j_1} & Z_{i_1,j_2} \\ Z_{i_2,j_1} & Z_{j_2,j_2} \end{pmatrix} = 0$ for all i_1, i_2 and j_1, j_2. We want to show that the morphism $S : \mathbf{P}^n \times \mathbf{P}^m \to T$ has an inverse which is a morphism. We define $S^{-1} : D(Z_{i,j}) \to D(X_i) \times D(Y_j)$ in local coordinates by $S^{-1}(z_{p,q}) = (z_{p,j}) \times (z_{i,q})$. So S^{-1} is clearly a morphism and one easily checks using the equations of T that S^{-1} is the inverse of S. $\qquad \square$

We get

Corollary 3.2.2. *The product of (quasi-)projective varieties is (quasi-) projective.*

As we don't have much to do in this section I will give the first example of the theorem.

$\mathbf{P}^1 \times \mathbf{P}^1$ is isomorphic to the rank 4 quadric $XW - YZ = 0$ in $\mathbf{P}^3$.

3.3 Graphs of morphisms and separatedness

Let $f : X \to Y$ be a morphism between spaces with functions. Consider $\Gamma_f : X \to X \times Y$, the morphism such that $\pi_X \circ \Gamma_f = 1_X$ and $\pi_Y \circ \Gamma_f = f$.

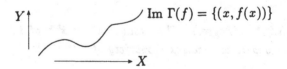

The image of Γ_f is $\mathrm{graph}(f)$ which is a subspace with functions. We have the morphism $\Gamma_f : X \to \mathrm{graph}(f)$ which is an isomorphism because its inverse is $\pi_X|_{\mathrm{graph}(f)}$.

The most common graph is the diagonal
$$\{(x,x)\}$$
$$\|$$
$$\Delta_X \subset X \times X$$
which is the graph(1_X). We have a trivial set-theoretic equation
(∗) $(f, 1_Y)^{-1}(\Delta_Y) = \text{graph}(f)$ where $(f, 1_Y) : X \times Y \to Y \times Y$ is the product morphism of f and the identity 1_Y of Y.

Exercise 3.3.1. Check equation ∗.

A variety X is *separated* if the diagonal Δ_X is a closed subset of the product $X \times X$. Varieties for many authors are assumed to be separated and even irreducible. So one must be careful to understand what any given author means by variety. Much of the theory does not use separatedness and one sometimes constructs varieties without knowing (*a priori*) that they are separated.

Lemma 3.3.2.

(a) *If $f : X \to Y$ is a morphism and Y is a separated variety then graph(f) is closed in $X \times Y$.*

(b) *A subvariety of a separated variety is separated.*

(c) *The product of two separated varieties is separated.*

(d) *(Quasi-)affine and (quasi-)projective varieties are separated.*

Proof. For (a) by (∗) graph(f) is closed if Δ_Y is closed. For (b) if $X \subset Y$ is a subvariety $\Delta_X = \Delta_Y \cap (X \times X)$. So Δ_X is closed if Δ_Y is. For (c) $\Delta_{X \times Y} = \pi_{1,3}^{-1} \Delta_X \cap \pi_{2,4}^{-1} \Delta_Y \subset X \times Y \times X \times Y$. (d) is also easy. The diagonal Δ_X of an affine variety X has equations $f(x_1) = f(x_2)$ for all f in $k[X]$. The quasi-case follows from (b). For the projective case we need only check that $\Delta_{\mathbf{P}^n}$ is closed in $\mathbf{P}^n \times \mathbf{P}^n \overset{S}{\hookrightarrow} \mathbf{P}^{n^2+2n}$. One simply checks that

(†) $\Delta_{\mathbf{P}^n} = \mathbf{P}^n \times \mathbf{P}^n \cap \{Z_{i,j} = Z_{j,i} \text{ for all } j \text{ and } i\}$.

$\square$

Exercise 3.3.3. Prove (†) (locally on $\mathbf{P}^{n^2+2n}$).

Lemma 3.3.4. *If $f : X \to Y$ is a morphism of two varieties then graph(f) is locally closed in $X \times Y$.*

Proof. Let $Y = \bigcup V_i$ be an open affine cover of Y. Let $f_i : f^{-1}V_i \to V_i$

be the induced morphism. Now $\text{graph}(f) = \bigcup \text{graph}(f_i)$ and $\text{graph}(f_i)$ is closed in the open $f^{-1}V_i \times V_i$ as the affine variety V_i is separated. Therefore $\text{graph}(f)$ is a closed subset of an open subset of $X \times Y$.   □

We will give an example of a variety which is not separated: a line with two zeroes. Let $X = (\mathbb{A}^1 = \{(x)\}) \amalg (\mathbb{A}^1 = \{(y)\})$ modulo $x \approx y$ if $x \neq 0 \neq y$ and $x = y$. Thus $X = \mathbb{A}^1 - \{0\} \amalg \{0_x\} \amalg \{0_y\}$. Let $\pi : \mathbb{A}^1 \amalg \mathbb{A}^1 \to X$ be the quotient mapping; we give X the quotient space with functions structure. Then one checks that $(\mathbb{A}^1 - \{0\} \amalg \{0_x\}) \bigcup (\mathbb{A}^1 - \{0\} \amalg \{0_y\})$ is an open affine covering of X. To see that X is not separated look at the morphism $\psi : \mathbb{A}^1 \to X \times X$ given by the morphism $\mathbb{A}^1 \to X$ and $\mathbb{A}^1 \to X$. Then $\psi(0) = (0_x, 0_y)$ but $\psi(\mathbb{A}^1 - \{0\}) \subset \Delta_X$. Thus $(0_x, 0_y)$ is contained in $\overline{\Delta}_X - \Delta_X$ but is not in Δ_X. So Δ_X is not closed.

We want to have an algebraic criterion for separatedness in terms of X.

Proposition 3.3.5.   *Let X be a variety. The following are equivalent:*

(a)   *X is separated;*

(b)   *for all open affine subsets U and V of X we have $U \cap V$ is affine and $k[U] \otimes_k k[V] \to k[U \cap V]$ is surjective.*

Proof.   Assume that X is separated. Then $U \cap V \approx \Delta_{U \cap V} = \Delta_X \cap (U \times V)$ is closed in the affine variety $U \times V$. Thus $U \cap V$ is affine and $k[U \times V] \to k[U \cap V]$ is surjective. Convesely (b) implies that $\Delta_X \cap (U \times V)$ is closed in $U \times V$ but $X \times X$ has an open cover by such affines. So δ_X is closed.   □

Exercise 3.3.6.   Let f and $g : X \to Y$ be two morphisms where Y is separated. Then $f = g$ if f and g agree on an open dense set.

Exercise 3.3.7.   In the above situation if $h : 0 \to Y$ is a morphism where 0 is a dense open subset of X, there is a maximal open subset of X to which h extends to a morphism.

3.4 Algebraic groups

A variety G is an *algebraic group* if we have morphisms $m : G \times G \to G$ and inv: $G \to G$ such that G is set-theoretically a group with multiplication m and inverse inv.

We will give a few examples.

(1)   The additive group $\mathbb{G}_a$ is $\mathbb{A}^1$ with addition $m(x,y) = x + y$ and $\text{inv}(x) = -x$.

(2)   The multiplication group $\mathbb{G}_m$ is $\mathbb{A}^1 - \{0\}$ with multiplication $m(x,y) = xy$ and $\text{inv}(x) = x^{-1}$.

(3)   The general linear group $\mathbb{GL}(n)$ = the affine space of $n \times n$ matrices with non-zero determinant. The group law is matrix multiplication and the inverse given by Cramer's rule is a polynomial of the coefficients and 1/determinant.

Not all algebraic groups are affine. We will later see some examples. For now we just note

Lemma 3.4.1.

An algebraic group is separated.

Proof. Let (g_1, g_2) be a point of the product. This is in the diagonal iff $g_1 g_2^{-1} = m(g_1, \text{inv } g_2) = e$ (the identity). As $\{e\}$ is closed, its inverse image Δ_X by a morphism is closed.   □

3.5 Cones and projective varieties

A *cone* in $\mathbb{A}^{n+1}$ with vertex 0 is a subset $C \ni 0$ such that λc is in C for all λ in k and c in C. Thus a cone is just the union of lines through the origin or just $\{0\}$. A cone C in $\mathbb{A}^{n+1}$ defines a subset $[C]$ in $\mathbb{P}^n$ and conversely. Here $[C] = \pi(C - \{0\})$ where π is the projection $\pi : \mathbb{A}^{n+1} - \{0\} \to \mathbb{P}^n$. By definition of the topology in $\mathbb{P}^n$, closed cones correspond to closed subsets of $\mathbb{P}^n$.

Lemma 3.5.1. *If the cone C is a subvariety of $\mathbb{A}^{n+1}, \dim[C] = \dim C - 1$.*

Proof.   Let $\pi : C - \{0\} \to [C]$ be the induced morphism. Let U be the open affine in $\mathbb{P}^n$ where the i-th coordinate is non-zero. The point is that we have an obvious isomorphism $\psi : ([C] \cap U) \times \mathbb{G}_m \to (C \cap \pi^{-1}(U))$, given by $\psi((c_0, \ldots, 1, \ldots, c_n), \lambda) = (c_0\lambda, \ldots, \lambda, \ldots, \lambda c_n)$. This is an isomorphism. Why?   □

Thus sometimes one thinks of a projective variety as being the lines in a cone in an affine space.

Let X be a closed subset in $\mathbb{P}^n$. Then the cone over X is a closed subset C of $\mathbb{A}^{n+1}$.

Claim. The ideal of C is generated by homogeneous polynomials.

Proof. Let f vanish on C. Then $\lambda^* f(x) = f(\lambda x)$ also vanishes on C. Thus λ^* gives an action of k^* on the ideal I of C. Thus I is spanned by the eigenvectors in I which are the homogeneous polynomials.   □

Thus $X = \text{zeroes}(f_1, \ldots, f_n)$ where the f_i are homogeneous polynomials in the homogeneous coordinates. In particular $D(f) = X - \text{zeroes}(f)$ are a basis of the topology of X where f is homogeneous.

Exercise 3.5.2. Show that $D(f)$ is affine if $f \neq 0$.

We will need to know

Theorem 3.5.3. (Projective nullstellensatz). *Let I be a homogeneous ideal in $k[\mathbf{A}^{n+1}]$. If f is a homogeneous polynomial such that $\text{zeroes}(f)$ in $\mathbf{P}^n$ contains the projective locus of $\text{Cone}(\text{zeroes}(I))$, then $f^d \in I$ for some $d > 0$. In particular $\text{zeroes}(I)$ in $\mathbf{P}^n$ is empty iff I contains all homogeneous forms of degree d for some d.*

Proof. $\text{Cone}(\text{zeroes}(I)) = (\text{ordinary-zeroes}(I))$. By assumption f vanishes on ordinary-zeroes$(I) - \{0\}$ but it vanishes at zero because it is homogeneous. Then the nullstellensatz gives $f^n \in I$ for $n > 0$. The second statement is a simple application with f equal to the coordinate functions $X_0, \ldots, X_n$.   □

3.6 A little more dimension theory

We begin with the affine case.

Theorem 3.6.1. *Let X and Y be two irreducible closed subvarieties of $\mathbf{A}^n$. If it is non-empty, each component of $X \cap Y$ has dimension $\geq \dim X + \dim Y - n$.*

The projective version is an existence theorem.

Theorem 3.6.2. *Replace $\mathbf{A}^n$ by $\mathbf{P}^n$, then if this number ≥ 0, $X \cap Y$ is non-empty and each component has dimension $\geq \dim X + \dim Y - n$.*

First we do

$$\text{Theorem } 3.6.1 \Longrightarrow \text{Theorem } 3.6.2.$$

Let $C(X)$ denote the cone in $\mathbf{A}^{n+1}$ corresponding to a projective variety X. Then the intersection $C(X) \cap C(Y) \ni 0$ is non-empty. Thus if Z is a component of $X \cap Y$ then $C(Z)$ is a component of $C(X) \cap C(Y)$. Thus $\dim Z = \dim C(Z) - 1 \geq (\dim C(X) + \dim C(Y) - (n+1)) - 1 = \dim X + \dim Y - n$. If the number is non-negative $\dim C(Z) \geq 1$. Thus $C(Z) \neq \{0\}$ and Z is non-empty. It remains to complete the

Proof of Theorem 3.6.1. The trick is called reduction to the diagonal. This uses the fact that $X \cap Y \approx \Delta_{X \cap Y} = (X \times Y) \cap \Delta_{\mathbf{A}^n}$. Now $\Delta_{\mathbf{A}^n} =$ zeroes$(x_1 - y_1, x_2 - y_2, \ldots, x_n - y_n)$ in $\mathbf{A}^n \times \mathbf{A}^n = \{(x)\} \times \{(y)\}$ is the zeroes of n functions. By the corollary of the principal ideal theorem a non-empty component of the intersection has dimension at least $\dim(X \times Y) - n = \dim X + \dim Y - n$.

3.7 Complete varieties

A variety X is *complete* if it is separated and for all varieties Y and any closed subset $Z \subset X \times Y$ the projection $\pi_Y(Z)$ is closed in Y.

Clearly closed subvarieties of complete varieties are complete. Also morphic images of complete varieties in separated varieties are complete.

Exercise 3.7.1. Prove these statements.

Lemma 3.7.2. *If X is an irreducible complete variety, then any regular function on X is a constant.*

Proof. Let g be a regular function on X. Assume that $g \not\equiv 0$. Then $D(g)$ is a dense open subset of X. Consider the graph $\Gamma \equiv (yg(x) = 1)$ of $\frac{1}{g}$ in $X \times \mathbf{A}^1$. Then $\pi_{\mathbf{A}^1}(\Gamma)$ does not contain 0 but as X is complete and irreducible it is a closed proper subset of $\mathbf{A}^1$. Therefore $\pi_{\mathbf{A}^1}(\Gamma)$ is one point, say $1/P$. Thus $g(x) = P$ if $g(x) \neq 0$. Thus $g \equiv P$. □

Exercise 3.7.3. Show that a quasi-affine complete variety is a finite set.

The main result is

Theorem 3.7.4. *Projective varieties are complete.*

Proof. By the previous remarks it will suffice to show that $\mathbf{P}^n$ is complete. Let $Z \subset Y \times \mathbf{P}^n$ be a closed subset where Y is a variety. As the problem of showing that $\pi_Y(Z)$ is closed is local on Y we may assume that Y is affine.

Claim. We may find homogeneous polynomials f_j in $k[Y][X_0, \ldots, X_n]$ such that $Z = \{y, (x_0, \ldots, x_n)) \in Y \times \mathbf{P}^n | f_i(y, x) = 0$ for all $i\}$.

Let $C(Z) = $ closure of $(1_Y \times \pi)^{-1}(Z)$ in $Y \times \mathbf{A}^{n+1}$. Then $C(Z)$ is stable under the action $\lambda_*(y, x) = (y, \lambda x)$ where $\lambda \in \mathbf{G}_m$. Thus the ideal I of $C(Z)$ is $\mathbf{G}_m$-invariant. Thus I has a basis f_j of eigenvectors which are homogeneous polynomials in the X_i. As $Y \times \mathbf{A}^{n+1}$ is affine, this solves the problem of the claim.

We will show that $\pi_Y(Z)$ is closed by writing it as the intersection of determinantal subvarieties. Let a_j be the degree of f_j in the X_i's. Let P_j be the vector space of all homogeneous polynomials in $k[X_0, \ldots, X_n]$ of degree a_j. Then consider the linear combinations $\sum P_* f_*$. They define a $k[Y]$-line mapping $\beta_n : \bigoplus_j P_{n-a_j} \otimes_k k[Y] \to P_n \otimes_k k[Y]$ for all n. Thus in terms of a basis of the P_*, β_n is given by a matrix α_n with coefficients in $k[Y]$. Let E_n be the determinantal subvariety $\{$rank $\alpha_n(y) \leq \dim P_n - 1\}$. Then y is in E_n if and only if $\beta_n(y)$ is not surjective.

The projective nullstellensatz says that $y \in \pi_Y(Z) \Leftrightarrow \pi_Y^{-1}(y) \neq \emptyset \Leftrightarrow f_1(y, X) = \ldots = f_n(y, X) = 0$ is a non-empty subset $\Leftrightarrow \beta_n(y)$ is not surjective for all $n \Leftrightarrow y \in \bigcap_n E_n$ for all n. Thus $\pi_Y(Z)$ is the intersections of closed determinantal subvarieties. Hence $\pi_Y(Z)$ is closed. $\square$

Remark. Using some known algebra it can be shown that $\pi_Y(z) = E_n$ where $n = \{(\max$ degree of $f_i) - 1\}(n + 1) + 1$.

3.8 Chow's lemma

Let $f : X \to Y$ be a morphism. Then f is *birational* if there are open dense subsets U and V of X and Y such that f induces an isomorphism $U \xrightarrow{\approx} V$. Thus X and Y are mostly the same.

Lemma 3.8.1. (Chow.) *Let Y be a complete irreducible variety. Then there exist a projective variety X and a birational morphism $f : X \to Y$.*

Proof. Actually we will prove a stronger result under just the assumption that Y is irreducible and separated. This stronger result is

(1) There exist an irreducible projective variety W and a closed subvariety $Z \subset W \times Y$ such that $\pi_Y : Z \to Y$ is birational and $\pi_W : Z \to W$ gives an isomorphism between Z and an open subset U of W.

First we note how this implies the lemma. If Y is complete, $\pi_W(Z) = U$ is closed. Hence $U = W$. Take $X = W$ and $f = \pi_Y \circ \pi_W^{-1}$. It remains to prove (1).

Let $Y = Y_1 \cup \ldots \cup Y_p$ be an affine cover of Y by open dense subsets. Find a projective variety $\overline{Y}_i$ which contains Y_i as an open subset. Let $O = Y_1 \cap \ldots \cap Y_p$. We have the inclusion $O \overset{\text{diagonal}}{\hookrightarrow} O \times \ldots \times O \hookrightarrow Y_1 \times \ldots \times Y_p \hookrightarrow \overline{Y}_1 \times \ldots \times \overline{Y}_p$ where the first inclusion is closed, the second is open, and the third open. Hence O is open in its closure $\overline{O}$ in $\overline{Y}_1 \times \ldots \times \overline{Y}_p$. Let $W = \overline{O}$ and let $Z =$ closure in $W \times Y$ of the graph of the inclusion of O in Y. Thus by construction $\pi_Y : Z \to Y$ is birational as it is the identity on O. It remains to check that $\pi_W : Z \to W$ is an isomorphism between Z and an open subset U of W.

Consider the open subset $Z_i = Z \cap (W \times Y_i)$ for some i. If we can show that Z_i is the graph of a morphism $\psi_i : U_i \to Y_i$ where U_i is an open subset of W, we will be done because the Z_i cover Z and $U = U_1 \cup \ldots \cup U_p$ is open in W. There is no secret what U_i is. It is $W \cap \overline{Y}_1 \times \ldots \times Y_i \times \ldots \times \overline{Y}_p$. The morphism ψ_i just reads the i-coordinate in the product. All we have to do is to show that if $(y_1, \ldots, y_p, y)$ is in Z_i then $y_i = y$. Thus we want to show that Z_i is contained in the closed subset $y_i = y$ of $U_i \times Y_i$. This is trivial because the transposed graph (y, y) of the inclusion Y_i in $\overline{Y}_i$ is closed in $\overline{Y}_i \times \overline{Y}_i$ and is the closure of $\{(u, u) | u \in O\}$. $\qquad\square$

3.9 The group law on an elliptic curve

An *abelian variety* is an irreducible complete algebraic group. These varieties have a special position in algebraic geometry as the study of geometry on a general variety sometimes leads one to study associated abelian varieties.

An abelian curve is frequently called an *elliptic curve*. In this section we want to study the group law on one elliptic curve.

Consider E, the closure in $\mathbf{P}^3$ of the curve $y^2 = x^3 + x$ in $\mathbf{A}^2$. Clearly this curve is irreducible and it is complete because it is projective. Thus $E = \text{zeroes}(ZY^2 - Z^2X - X^3)$. Now $E \cap$ line at $\infty =$ zeroes $\{ZY^2 - Z^2X - X^3, Z\}$ is the sole point $(0, 0, 1)$ in (z, x, y). This point will be the zero 0 of our group. The inverse is the extension to E of the mapping $(x, y) \to (x, -y)$ on the affine curve. This extends trivially to E in projective coordinates $(x, y, z) \to (x, -y, z)$. Now addition $E \times E \to E$ is hard to define.

The geometry is simple. If e_1 and e_2 are distinct points of E, then $e_1 + e_2 = -(e_3)$ where e_3 is the other point of $E\cap$ line spanned by e_1, e_2:

We will try to compute $e_3 = (x_3, y_3)$ in affine coordinates. The line spanned by $e_1 = (x_1, y_1)$ and $e_2 = (x_2, y_2)$ is given by $\{(x_1, y_1) + \lambda[(x_2, y_2) - (x_1, y_1)] = P(\lambda)$ for λ in $k\}$. The equation for λ such that $P(\lambda)$ is on E is $((1-\lambda)y_1 + \lambda y_2)^2 - ((1-\lambda)x_1 + \lambda x_2)^3 - ((1-\lambda)x_1 + \lambda x_2) = 0 = \lambda(1-\lambda)f(\lambda, x_1, x_2, y_1, y_2)$ where f is linear in λ. Solve $f = 0$ for λ in terms of x_1, x_2, y_1, y_2. Then the formulas for x_3 and y_3 are rational functions of x_1, x_2, y_1, y_2. If you work hard enough you can show that the addition extends to a morphism $E \times E \to E$ and the associative, commutative, and inverse rules hold. We don't have to learn these calculations because our theory of curves will eventually prove all these facts for all smooth plane cubics. Newton discovered these groups but it took years to understand them theoretically without gross calculation.

3.10 Blown up $\mathbf{A}^n$ at the origin

There exist many interesting birational morphisms. In this section we will give an interesting example of a special case of a *blown up* affine space.

Let $\mathbf{A}^n$ be affine space. Then we may consider $\mathbf{P}^{n-1}$ as consisting of lines (through 0) in $\mathbf{A}^n$. Thus a point in $\mathbf{A}^n \times \mathbf{P}^{n-1}$ is a pair (p, ℓ) of a point p and a line ℓ in $\mathbf{A}^n$. Consider $Z \equiv \{(p, \ell)|p \in \ell\} \subset \mathbf{A}^n \times \mathbf{P}^{n-1}$.

Thus Z has two projections $\pi_{\mathbf{P}^{n-1}} : Z \to \mathbf{P}^{n-1}$ and $\pi_{\mathbf{A}^n} : Z \to \mathbf{A}^n$. What we want to prove is

Exercise 3.10.1.

(a) Z is a closed subvariety of $\mathbf{A}^n \times \mathbf{P}^{n-1}$.

(b) $\pi_{\mathbf{A}^n}$ is birational and an isomorphism over $\mathbf{A}^n - \{0\}$.

(c) $\pi_{\mathbf{P}^{n-1}}$ is a locally trivial bundle of lines.

(d)   The exceptional divisor $E \equiv \pi_{\mathbf{A}^n}^{-1}(0)$ is locally defined by one equation in Z. (Hint: Where is E in the line bundle of (c)?)

4

Sheaves

Let's face it. You can not read most of the current literature in algebraic geometry without knowing *sheaf* theory. *Leray* developed his original version of sheaf theory as a means of analyzing the global obstructions to piecing together local data. Previously, many geometers and topologists had worked on these kinds of problems. At the present time, the previous and current work is normally expressed in the language of sheaf theory.

4.1 The definition of presheaves and sheaves

Let X be a topological space. A *presheaf* F on X is the assignment of a set $F(V)$ to any open subset V of X together with a restriction mapping $\operatorname{res}_U^V : F(V) \to F(U)$ whenever $U \subset V$ are two open subsets of X such that

(*a*) $\operatorname{res}_U^U = $ identity of $F(U)$ for all open subsets U, and

(*b*) $\operatorname{res}_U^V \circ \operatorname{res}_V^W = \operatorname{res}_U^W$ whenever $U \subset V \subset W$ are three open subsets of X.

A *morphism* $\alpha : F \to G$ between two presheaves on X is a collection of mappings $\alpha(V) : F(V) \to G(V)$ for each open subset V of X such that the diagram

$$(\star) \qquad \begin{array}{ccc} F(V) & \xrightarrow{\ \alpha(V)\ } & G(V) \\ \text{res}_U^V \downarrow & & \downarrow \text{res}_U^V \\ F(U) & \xrightarrow{\ \alpha(U)\ } & G(U) \end{array}$$

commutes whenever $U \subset V$ are two open subsets of X.

First, we will develop a more useful notation for presheaves. If F is a presheaf on X and V is an open subset of X, then an element σ of $F(V)$ is called a *section* of F over V. If U is an open subset of V, the section $\text{res}_U^V(\sigma)$ over U is denoted by $\sigma|_U$ and is called the *restriction* of σ to U. If we have a morphism $\alpha : F \to G$ of two presheaves and σ is a section of F over V, then $\alpha(\sigma) \equiv \alpha(V)(\sigma)$ is a section of G over V.

In this improved notation, we have equations expressing the above lines of the same names.

(a) $\sigma|_U = \sigma$ for all sections σ of F over U.

(b) $(\sigma|_V)|_U = \sigma|_U$ for all sections σ of F over W.

$(\star)$ $\qquad (\alpha(\sigma))|_U = \alpha(\sigma|_U)$ for any section σ of F over V.

One method of studying a presheaf F is to examine the behaviour of sections of F over smaller and smaller neighborhoods of a point x of X. Formally, the *stalk* F_x of F at the point x is $\varinjlim_{V \ni x} F(V)$, where the V's are open neighborhoods of x, which are partially ordered by inclusion. Informally, F_x is the disjoint union $\bigcup F(V)$ of all sections of F over some open neighborhood of x modulo two sections being equivalent if they have the same restriction to a neighborhood contained in their domains of definition.

If σ is a section of F over some neighborhood of x, σ defines an element σ_x in the stalk F_x of F at x. Here, σ_x describes the behavior of σ "arbitrarily near to x". We will next give a lemma which says that stalks are natural.

Lemma 4.1.1.

Let $\alpha : F \to G$ be a morphism of presheaves on X.

(a)   Then α induces a unique mapping $\alpha_x : F_x \to G_x$ such that $\alpha_x(\sigma_x) = (\alpha(\sigma))_x$ for any section σ of F over an open neighborhood of x.

(b)   (identity of $F)_x = $ identity of F_x.

(c)   If $\beta : G \to H$ is another morphism of presheaves on $X, (\beta \circ \alpha)_x = (\beta_x) \circ (\alpha_x)$.

Proof. For (*a*), the uniqueness is obvious because any element of F_x may be written as σ_x for some section of F near x. To see the existence, one needs to prove that if σ and τ are sections of F near x such that $\sigma_x = \tau_x$, then $(\alpha(\sigma))_x = (\alpha(\tau))_x$. If $\sigma_x = \tau_x$, there is a neighborhood W such that $\sigma|_W$ and $\tau|_W$ are defined and equal. Thus $(\alpha(\sigma)|_W) = \alpha(\sigma|_W) = \alpha(\tau|_W) = (\alpha(\tau))|_W$. Hence, $\alpha(\sigma)_x = \alpha(\tau)_x$. This proves (*a*). Parts (*b*) and (*c*) are obvious. $\qquad\square$

In general, one loses information about a presheaf by just looking at its stalks. To analyse this loss, one introduces the presheaf $D(F)$ of "*discontinuous sections*" of F. For any open subset V of X, define $D(F)(V) = \prod_{v \in V} F_v$. Thus, a section τ of $D(F)$ over V is a collection $(\tau_v)_{v \in V}$, where τ_v is an element of the stalk F_v. The restriction $\operatorname{res}_U^V : D(F)(V) \to D(F)(U)$ for an open subset U of V sends $(\tau_v)_{v \in V}$ to $(\tau_u)_{u \in U}$. Clearly, $D(F)$ is a presheaf.

Let σ be a section of F over an open subset U of X. Then σ defines a section $i(\sigma)$ of $D(F)$ over U by the formula $i(\sigma) = (\sigma_u)_{u \in U}$. Thus, $i(\sigma)$ describes the behavior of σ "arbitrarily near" every point of U. Some additional facts about these constructions are contained in

Lemma 4.1.2.

(*a*) $i : F \to D(F)$ *is a morphism of presheaves.*

(*b*)   *If* $\alpha : F \to G$ *is a morphism of presheaves, there is a unique morphism* $D(\alpha) : D(F) \to D(G)$ *such that the diagram*

$$
\begin{array}{ccc}
\alpha : & F & \longrightarrow & G \\
 & \downarrow{\scriptstyle i} & & \downarrow{\scriptstyle i} \\
D(\alpha) : & D(F) & \longrightarrow & D(G)
\end{array}
$$

commutes.

(*c*) $D(\text{identity of } F) = \text{identity of } D(F)$.

(*d*)   *If* $\beta : G \to H$ *is another morphism of presheaves,* $D(\beta \circ \alpha) = D(\beta) \circ D(\alpha)$.

Proof. Let σ be a section of F over an open subset V that contains an open subset U. Part a) asserts that $i(\sigma|_U) = i(\sigma)|_U$. This means that $(\sigma|_U)_u = \sigma_u$ for any point u of U. Thus, (*a*) is true as the equation is evident.

For (*b*), let $\tau = (\tau_v)_{v \in V}$ be a section of $D(F)$ over V. Define $D(\alpha)(\tau) = (\alpha_v(\tau_v))_{v \in V}$, which is a section of $D(G)$ over V. By Lemma 5.1.1(*a*), $D(\alpha)$ is the only possibility such that the diagram commutes, because commuting means that $D(\alpha)((\sigma_v)_{v \in V}) = (\alpha(\sigma)_v)_{v \in V}$ for any section σ of F over any V. It remains to check that $D(\alpha)$ is a morphism. Let U be an open subset of V. We need to see that, for any section $\tau = (\tau_v)_{v \in V}$, $D(\alpha)(\tau|_U) = (D(\alpha)(\tau))|_U$, but $D(\alpha)(\tau|_U) = D(\alpha)((\tau_u)_{u \in U}) = (\alpha_u(\tau_u))_{u \in U} = (\alpha_v(\tau_v))_{v \in V}|_U = (D(\alpha)(\tau))|_U$. This proves (b).

The proofs of (c) and (d) are trivial. $\qquad\square$

The exact amount of information that is lost by passing to stalks is described in

Lemma 4.1.3. *Let σ and τ be two sections of a presheaf F over an open subset V of X. Then $\sigma_v = \tau_v$ for all points v of V if and only if there is an open covering $V = \bigcup V_\alpha$ such that $\sigma|_{V_\alpha} = \tau|_{V_\alpha}$ for each α.*

Proof. It is more instructive to do this yourself. $\qquad\square$

A presheaf F is called *decent* if any section of F is determined by its local behavior. Thus, F is decent if, for any two sections σ and τ over an open subset V, $\sigma = \tau \Leftrightarrow \sigma_v = \tau_v$ for all v in $V \Leftrightarrow$ for some covering $V = \bigcup V_\alpha$, $\sigma|_{V_\alpha} = \tau|_{V_\alpha}$ for all α. We now proceed on to presheaves that are better than decent ones.

Another method for studying a presheaf F is to examine to what extent one may piece together local sections. Let $V = \bigcup V_\alpha$ be an open covering of an open subset V of X. For any section σ of F over V, we have a section $\sigma_\alpha \equiv \sigma|_{V_\alpha}$ of F over V_α for all α. This collection $\{\sigma_\alpha \in F(V_\alpha)\}$ is not arbitrary. In fact, it must satisfy the *patching condition*:

$$(\dagger) \qquad \sigma_\alpha|_{V_\alpha \cap V_\beta} = \sigma_\beta|_{V_\alpha \cap V_\beta} \quad \text{for all pairs } \alpha \text{ and } \beta.$$

This is because $\sigma_\alpha|_{V_\alpha \cap V_\beta} = (\sigma|_{V_\alpha})|_{V_\alpha \cap V_\beta} = \sigma|_{V_\alpha \cap V_\beta} = (\sigma|_{V_\beta})_{V_\alpha \cap V_\beta} = \sigma_\beta|_{V_\alpha \cap V_\beta}$.

The more than decent presheaves are called *sheaves*. Basically, a sheaf is a presheaf whose sections are locally determined and closed under piecing together. Formally, a sheaf F is a decent presheaf such that, for any family V_α of open subsets of X and any collection $\{\sigma_\alpha \in F(V_\alpha)\}$ of sections of F over each V_α that satisfies the patching condition $(\dagger)$, there is a section σ of F over the union $\bigcup V_\alpha$ such that $\sigma_\alpha = \sigma|_{V_\alpha}$ for each α. The decency of F is equivalent to the uniqueness of the section σ gotten by piecing the σ_α's together.

At first sight, the definition of a sheaf seems complicated. As sheaves frequently arise in mathematics, there is great economy in learning to think of a sheaf as one idea and not writing out all these conditions every time that you deal with them. Actually, we have been dealing with sheaves for sometime as the next example will show.

Example. Let X be a space with functions. The assignment $V \mapsto k[V]$ is a sheaf, where the restrictions are given by restriction of functions.

Example. One often deals with sheaves of functions that satisfy some local condition; for instance, solutions of a differential equation. One may work with sheaves of differentials.

Example. Let F be a presheaf on a topological space. The presheaf $D(F)$ of "discontinuous sections" of F is a sheaf. It is easy to see that "discontinuous sections" are locally determined and may be pieced together from local data satisfying the patching conditions.

Exercise 4.1.4. Give an example of a presheaf that is not decent.

Exercise 4.1.5. Give an example of a decent presheaf that is not a sheaf.

Exercise 4.1.6. Let $F(U) = \{$all complex analytic functions f on $U | z\frac{df}{dz} = 1\}$, for any domain U in $\mathbb{C}$. Show that F is a sheaf under restriction of functions.

Exercise 4.1.7. In Exercise 4.1.6, prove that the stalk of F at the point O is empty and its stalk at any other point is non-canonically isomorphic to $\mathbb{C}$.

Exercise 4.1.8. Let σ and τ be sections of a decent presheaf over an open subset U. Show that the subset $\{u \in U | \sigma_u = \tau_u\} = V$ is open.

Exercise 4.1.9. What sheaves do you know for which the subset V is always closed in U?

4.2 The construction of sheaves

Let X be a topological space. Given two (pre-)sheaves F and G on X, F is called a *sub-(pre-)sheaf* of G, if, for all open subsets V of X, $F(V)$ is a subset of $G(V)$ and the restriction in F coincides with the restriction in G. Thus, the inclusion $F \subset G$ gives a morphism of (pre-)- sheaves.

Example. If X is a space with functions, then the sheaf $V \mapsto k[V]$ is a subsheaf of the sheaf of all k-valued functions $V \mapsto k^V$ with restriction of functions. The sheaf $V \mapsto k[V]$ is called the *structure sheaf* or sheaf of regular functions on X. The structure sheaf is usually denoted by $\mathcal{O}_X$.

We will next explain some useful facts about sub-(pre-)sheaves.

Lemma 4.2.1. *Let x be a point of X. Let F and G be two presheaves on X.*

(a) *If F is a sub-presheaf of G, then the stalk F_x is a subset of G_x.*

(b) *A sub-presheaf of a decent presheaf is itself decent.*

(c) *If F is a sub-presheaf of a sheaf G, then there is a smallest sub-sheaf H of G, that contains F. Furthermore, the stalks F_x and H_x are equal.*

(d) *Let $\alpha : F \to G$ be a morphism of presheaves. Set pre-$\alpha(F)(V) =$ image of $F(V)$ in $G(V)$ via $\alpha(V)$ for all open subsets V of X. Then, pre-$\alpha(F)$ is a sub-presheaf of G and the image of the stalk mapping $\alpha_x : F_x \to G_x$ is (pre-$\alpha(F))_x$.*

Proof.

(a) Let σ and τ be two sections of F over a neighborhood of x. If their images in G_x are the same, their restrictions to a smaller neighborhood V at x agree in $G(V)$. As $F(V) \subset G(V)$, their restrictions agree in $F(V)$. Hence, σ and τ have the same image in F_x. Therefore, the stalk mapping $F_x \to G_x$ is injective.

(b) If G is decent, then the mapping $G(V) \to D(G)(V) = \prod_{v \in V} G_v$ is injective for all open subsets V of X. By (a), we must have an injection $F(V) \hookrightarrow D(F)(V) = \prod_{v \in V} F_v$. Hence, F is decent as its sections are locally determined.

(c) Let V be an open subset of X. As H is supposed to be a subsheaf of G, $H(V)$ must contain all sections of G that are gotten by piecing together sections of F. Define $H(V)$ to be the subset $\{\sigma \in G(V) | \sigma|_{V_\alpha} \in F(V_\alpha)$ for all α, where $V = \bigcup V_\alpha$ is some open covering of $V\}$. If we can show that H is a subsheaf of G, H will be the smallest such subsheaf and the stalk F_x will be clearly equal to H_x (because, if $x \in V$, then $x \in V_\alpha$ for some α).

To show that H is a sub-presheaf of G, let U be an open subset of V. We need to show that restriction in G takes $H(V)$ into $H(U)$. Let σ be a typical section of H for an open covering $V = \bigcup V_\alpha$. Then, $U = \bigcup(U \cap V_\alpha)$ is an open covering and $(\sigma|_U)|_{U \cap V_\alpha} = (\sigma|_{V_\alpha})|_{U \cap V_\alpha} \in$

$F(U \cap V_\alpha)$ as F is a sub-presheaf of G. Therefore, $\sigma|_U$ is contained in $H(U)$.

By (b), H is decent. It remains to prove that we may piece together sections of H. Let $W = \bigcup W_\beta$ be an open covering. Given a section σ_β of H over W_β satisfying the patching conditions, there exists a unique section σ of G over W such that $\sigma|_{W_\beta} = \sigma_\beta$ for each β because G is a sheaf. To prove that H is a sheaf, we have to see that $\sigma \in H(W)$. This follows from the definition of H, as the union of open coverings of each W_β is an open covering of W.

(d) is very easy. You can try to prove it for practice. □

We will next use the above idea to see that there is a natural way to modify a given presheaf to make it decent and, even, make it a sheaf.

Let F be a presheaf on X. Let $F^\flat$ be the sub-presheaf pre-$i(F)$ of $D(F)$, where $i : F \to D(F)$ is the morphism sending a section σ of F to the discontinuous section $(\sigma_v)_{v \in V}$. Let $F^\sharp$ be the smallest subsheaf of $D(F)$ that contains $F^\flat$. A section of $F^\sharp$ over an open subset U is a collection $(g_u)_{u \in U}$ where $g_u \in F_u$ such that there is an open covering $U = \bigcup U_i$ such that there exists sections f_i in $F(U_i)$ for each i such that $f_{i,u} = g_u$ for all u in U_i. The next lemma will basically record that these constructions are natural.

Lemma 4.2.2.

(a) We have morphisms of presheaves, $F \to F^\flat \subset F^\sharp \subset D(F)$. For any point x of X, the induced mappings $F_x \overset{\approx}{\to} F_x^\flat \overset{\approx}{\to} F_x^\sharp$ are isomorphisms. The presheaf $F^\flat$ is decent and $F^\sharp$ is a sheaf.

(b) If $\alpha : F \to G$ is a morphism of presheaves, we have unique morphsms, $\alpha^\flat : F^\flat \to G^\flat$ and $\alpha^\sharp : F^\sharp \to G^\sharp$, such that the diagram

$$
\begin{array}{ccccccc}
F & \longrightarrow & F^\flat & \subset & F^\sharp & \subset & D(F) \\
\downarrow{\scriptstyle\alpha} & & \downarrow{\scriptstyle\alpha^\flat} & & \downarrow{\scriptstyle\alpha^\sharp} & & \downarrow{\scriptstyle D(\alpha)} \\
G & \longrightarrow & G^\flat & \subset & G^\sharp & \subset & D(G)
\end{array}
$$

commutes.

(c) (identity of $F)^\flat$=identity of F and (identity of $F)^\sharp$=identity of $F^\sharp$.

(d) If $\beta : G \to H$ is another morphism of presheaves $(\beta \circ \alpha)^\flat = \beta^\flat \circ \alpha^\flat$ and $(\beta \circ \alpha)^\sharp = \beta^\sharp \circ \alpha^\sharp$.

Proof. All of the work for proving these statements has been done in the preceeding sequence of lemmas. □

In practice, one is given a presheaf F and one wants to prove that F is a sheaf, or, at least, differs negligibly from its sheafification $F^\sharp$. The exact result that we will need is

Lemma 4.2.3. *Let F be a presheaf on a topological space X. Let γ be a basis of open subsets of X. Assume that, for any open subset V in γ and any covering $V = \bigcup V_\alpha$ by members V_α of γ, if σ_α is a section of F over V_α for each α such that $\sigma_\alpha|_{V_\alpha \cap V_\beta} = \sigma_\beta|_{V_\alpha \cap V_\beta}$ for each pair α and β, there exists a unique section σ of F over V with $\sigma|_{V_\alpha} = \sigma_\alpha$ for each α. Then, for any open subset W in γ, the natural mappings $F(W) \to F^\flat(W) \to F^\sharp(W)$ are isomorphisms.*

Proof. Let W be an open subset in the basis γ. First, we will see that the surjection $F(W) \to F^\flat(W)$ is also injective and, hence, an isomorphism. Let σ and τ be two sections of F over W. Let $W = \bigcup W_\beta$ be an arbitrary covering of W by open subsets W_β. We need to see that $\sigma = \tau$ if $\sigma|_{W_\beta} = \tau|_{W_\beta}$ for each γ. As γ is a basis, we may refine the covering $W = \bigcup W_\beta$ by a covering $W = \bigcup V_\alpha$ where the V_α's are members of γ. Thus, $\sigma|_{V_\alpha} = \tau|_{V_\alpha}$ for all α and, by the uniqueness assumption, $\sigma = \tau$.

It remains to show that the composed injection $F(W) \overset{\cong}{\to} F^\flat(W) \hookrightarrow F^\sharp(W)$ is surjective. We know that any section σ of $F^\sharp(W)$ is gotten by piecing together sections of $F^\flat(W_\beta)$ for some covering $W = \bigcup W_\beta$. If we refine this covering to a covering $W = \bigcup V_\alpha$ by members V_α of γ, we see that σ is pieced together from sections of $F^\flat(V_\alpha)$. By the first part, $F(V_\alpha) \overset{\cong}{\to} F^\flat(V_\alpha)$ as $V_\alpha \in \gamma$. The existence part of the assumption now shows that σ is actually a section of F over W. Thus, $F(W) \overset{\cong}{\to} F^\sharp(W)$, whenever $W \in \gamma$. $\square$

Exercise 4.2.4. Let $F(U) = k$ for any open subset U of X, where the restrictions are the identity. Show that $F(U) \overset{\cong}{\to} F^\flat(U)$ may be identified with constant k-valued functions on U. Show that $F^\sharp(U)$ may be identified with the k-valued functions on U that are constant on the connected components of U if X is a locally connected topological space.

Exercise 4.2.5. Show that, if $\alpha : F \to G$ is a morphism from an arbitrary presheaf F to a decent presheaf G, then α factors uniquely as $F \to F^\flat \overset{\beta}{\to} G$.

Exercise 4.2.6. Show that, if $\alpha : F \to G$ is a morphism from an arbitrary presheaf F to a sheaf G, then α factors uniquely as $F \to F^\sharp \to G$.

4.3 Abelian sheaves and flabby sheaves

In this section, we discuss two special kinds of sheaves. We will be preparing the terrain for the next section, where these two kinds of sheaves will cooperate to give the cohomology of abelian sheaves constructed via Godement's canonical flabby resolutions.

A basic problem in sheaf theory concerns the determination of the sections of a given sheaf. In practice, one has a very good description of the sections of the sheaf over small open subsets and the problem is to understand the sections of the sheaf over large open subsets.

As one is dealing with a sheaf, one knows that the sections over a large open subset are gotten by piecing together local sections. I want to draw your attention away from the combinatorial aspects of "piecing together" approaches to attacking this problem. The function theorist more naturally asks the following question. "Given a section of a sheaf over an open subset, when does it extend to a section over a still larger open subset?"

A sheaf is called *flabby* if one may always solve this extension problem. More precisely, sheaf F on X is flabby if, for any open subset V of X, the restriction mapping $F(X) \to F(V)$ is surjective.

Example. Let G be any presheaf on a space X. Then the sheaf $D(G)$ of "discontinuous sections" of G is always a flabby sheaf.

We still can't say much about our extension problem unless we have a systematic way to measure the difference between two sheaves, say G and the flabby sheaf $D(G)$. If we introduce more structure on our sheaves, we will be able to say a lot about the extension problem.

An *abelian sheaf* (or sheaf of abelian groups) is a sheaf F on a space X, such that, for any open subset V of X, the set of sections $F(V)$ is an abelian group and the restriction $F(V) \to F(U)$ is a homomorphism of abelian groups for each open subset U of V. If one replaces "sheaf" by "presheaf", one also has a concept of an abelian presheaf.

Example. The structure sheaf $\mathcal{O}_X$ of a space with functions is an abelian sheaf. The group law is simply given by addition of functions.

The natural way to compare two abelian (pre-)sheaves is called a homomorphism. A *homomorphism* $\alpha : F \to G$ between two abelian (pre-)sheaves is a morphism such that $\alpha(V) : F(V) \to G(V)$ is a homomorphism of abelian groups for any open subset V.

The next lemma will assert that the constructions of the last two sections lead to abelian (pre-)sheaves if one starts from an abelian presheaf.

The only point of the omitted proof is that addition of sections commutes with restrictions.

Lemma 4.3.1. *Let F be an abelian presheaf on a space X with a point x.*

(a) *The stalk F_x admits a natural structure of an abelian group so that, for any open neighborhood V of x, the mapping $F(V) \to F_x$ is a homomorphism of groups.*

(b) *The decent presheaf F^b and the sheaves $F^\sharp$ and $D(F)$ are naturally abelian and the morphisms $F \to F^b \to F^\sharp \to D(F)$ are homomorphisms.*

(c) *For any homomorphism $\alpha : F \to G$ of abelian presheaves, the induced morphisms $\alpha_x, \alpha^b, \alpha^\sharp$ and $D(\alpha)$ are all homomorphisms and pre-$\alpha(F)$ is an abelian sub-presheaf of G.*

Next we will define exact sequences of abelian sheaves. Exactness gives a precise measurement of the difference between sheaves. Let 0 denote the abelian sheaf with only one section over any open subset.

A *short exact sequence* of abelian sheaves is a sequence of homomorphisms,

$$(*) \qquad 0 \to F_1 \overset{\alpha}{\to} F_2 \overset{\beta}{\to} F_3 \to 0,$$

between abelian sheaves on X such that, for any open subset V of X,

(a) the sequence of groups, $0 \to F_1(V) \to F_2(V) \to F_3(V)$, is exact, and

(b) F_3 is the smallest subsheaf containing pre-$\beta(F_2)$.

The last condition (b) means that any section of F_3 may be gotten by pasting together local sections of the form $\beta(\text{section of } F_2)$. A slick way to express the exactness of $(*)$ is to say that, for all points x of X, the sequence of stalks,

$$0 \to F_{1,x} \to F_{2,x} \to F_{3,x} \to 0,$$

is exact. The reader should check that *stalk exactness* is equivalent to the conditions (a) and (b).

One may try to use a short exact sequence, $0 \to F_1 \to F_2 \to F_3 \to 0$, to find out information about the sections of the big sheaf F_2 in terms of the sections of the smaller sheaves, F_1 and F_3. The main question, when you are trying to use this approach, is, "What is the image of $F_2(V)$ in $F_3(V)$ in the short exact sequence, $0 \to F_1(V) \to F_2(V) \to F_3(V)$?"

We now will see that this question is easily answered when F_1 is flabby.

Lemma 4.3.2.

Let $0 \to F_1 \to F_2 \to F_3 \to 0$ *be an exact sequence of abelian sheaves on a space* X.

(a) *If* F_1 *is flabby, then the sequence*

$$0 \to F_1(V) \to F_2(V) \to F_3(V) \to 0$$

is exact for any open subset V *of* X.

(b) *If both* F_1 *and* F_2 *are flabby, then* F_3 *is also flabby.*

Proof. First, we will see how (a) implies (b). We want to show that the restriction $t : F_3(X) \to F_3(V)$ is surjective for any open subset V of X. We have a commutative diagram,

$$
\begin{array}{ccc}
F_2(X) & \longrightarrow & F_3(X) \\
\downarrow{\scriptstyle s} & & \downarrow{\scriptstyle t} \\
F_2(V) & \xrightarrow{\; r \;} & F_2(V)
\end{array}
$$

of restrictions. By (a), r is surjective, and s is surjective because F_2 is flabby. Hence, t must be surjective.

For (a), we need to show that any section σ of F_3 over V lifts to a section of F_2 over V. Let τ_U be a lifting of $\sigma|_U$, where U is an open proper subset of V. If we can find an extension $\tau_{U'}$ of τ_U to a section of F_2 over a strictly larger open subset U' in V such that the extension $\tau_{U'}$ is a lifting of $\sigma|_{U'}$, then we may keep extending τ_U until we get a lifting of σ over all of V.

As U is a proper subset of V and our sequence is exact, we may find an open neighborhood W of a point in $V - U$ such that $\sigma|_W$ lifts to a section r_W of F_2 over W. If $r_W|_{U \cap W} = \tau_U|_{U \cap W}$, we may piece r_W and τ_U together to get a section $\tau_{U \cap W}$, which lifts $\sigma_{U \cup W}$. Thus, we have solved our problem with $U' = U \cup W$ as the patching condition is verified.

To finish the proof, we will see how to modify the choice of r_W to satisfy the patching condition. Let ρ be the difference $\tau_U|_{U \cap W} - r_W|_{U \cap W}$. Now, ρ is lifting of $\sigma|_{U \cap W} - \sigma|_{U \cap W} = 0$ to a section of F_2 over $U \cap W$. By the exactness, ρ comes from a section of F_1 over $U \cap W$. As F_1 is flabby, we may extend ρ to a section ρ' of F_2 over X which comes from a section of F_1 over X. Let $r'_W = r_W + \rho'|_W$. One immediately checks that r'_W and τ_U now satisfy the patching condition. $\qquad\square$

We will finish this section with a result showing the existence of some short exact sequences. Let F be an abelian subsheaf of an abelian sheaf

G on X. Let $U \subset V$ be two open subsets of X. We have a commutative diagram,

$$
\begin{array}{ccccccccc}
0 & \longrightarrow & F(V) & \hookrightarrow & G(V) & \xrightarrow{q(V)} & G(V)/F(V) & \longrightarrow & 0 \\
& & \downarrow{\scriptstyle \mathrm{res}^V_U} & & \downarrow{\scriptstyle \mathrm{res}^V_U} & & \downarrow{\scriptstyle K^V_U} & & \\
0 & \longrightarrow & F(U) & \hookrightarrow & G(U) & \xrightarrow{q(U)} & G(U)/F(U) & \longrightarrow & 0
\end{array}
$$

(+)

of homomorphisms of abelian groups.

Lemma 4.3.3.

(a) The assignment $V \to G(V)/F(V)$ with the restrictions $K_{\underset{U}{V}}$ is a presheaf, say pre-(G/F). The family $\{q(V)\}$ gives a homomorphism pre-$q : G \to$ pre-(G/F).

(b) Let $G/F = ($pre-$(G/F))^\sharp$. Then we have a natural short exact sequence of sheaves, $0 \to F \hookrightarrow G \xrightarrow{q} G/F \to 0$.

Proof. The presheaf axioms for pre-(G/F) follow from those of G. Thus, (a) is completely trivial. For (b), define q to be the composition $G \xrightarrow{\mathrm{pre}-q}$ pre-$(G/F) \to ($pre-$(G/F))^\sharp \equiv G/F$. To show the sequence in (b) is exact, we need only show that the sequence of stalks, $0 \to F_x \to G_x \to (G/F)_x \to 0$, is exact. By Lemma 4.2.2(a), (pre-$(G/F))_x \xrightarrow{\approx} ($pre-$(G/F))^\sharp_x \equiv (G/F)_x$. As the direct limit over $V \ni x$ of the exact sequences

$$0 \to F(V) \to G(V) \to G(V)/F(V) \to 0$$

is exact, the sequence $0 \to F_x \to G_x \to ($pre-$(G/F))_x \to 0$ is exact. Putting these two facts together, one gets the desired result. $\square$

Exercise 4.3.4. Show that the presheaf pre-(G/F) is always decent.

Exercise 4.3.5. Let C^∞ be the sheaf of real-valued C^∞-functions on $\mathbf{R}^2$. Let $\Omega(U)$ be the set of differentials $a(x,y)dx + b(x,y)dy$, where a and b are C^∞-functions defined on an open subset U of $\mathbf{R}^2$. Set $df = \frac{\partial f}{\partial x} \cdot dx + \frac{\partial f}{\partial y} \cdot dy$ for any C^∞-functions f. Show

(a) d defines a homomorphism $d : C^\infty \to \Omega$ of abelian sheaves on $\mathbf{R}^2$.

(b) Let $I\Omega$ be the smallest subsheaf of Ω that contains pre-$d(C^\infty)$. Explain why $I\Omega$ consists of "closed" (locally integrable) forms.

(c)   We have a short exact sequence of abelian sheaves,

$$0 \to \mathbf{R}^{\sharp} \to C^{\infty} \to I\Omega \to 0,$$

where $\mathbf{R}^{\sharp}$ is the sheaf of locally constant functions.

(d)   Give an example of a plane domain U such that $C^{\infty}(U) \to I\Omega(U)$ is not surjective.

Exercise 4.3.6.   Let

$$
\begin{array}{ccc}
F & \subset & G \\
\downarrow & & \downarrow \\
F' & \subset & G'
\end{array}
$$

be a commutative diagram of abelian sheaves. Show that one may naturally extend the diagram as

$$
\begin{array}{ccccccc}
F & \subset & G & \longrightarrow \text{pre-}(G/F) & \longrightarrow & G/F \\
\downarrow & & \downarrow & \downarrow & & \downarrow \\
F' & \subset & G' & \longrightarrow \text{pre-}(G'/F') & \longrightarrow & G'/F'
\end{array}
$$

Exercise 4.3.7.   If we have an exact commutative diagram of abelian sheaves,

$$
\begin{array}{ccccccccc}
0 & \longrightarrow & F_1 & \longrightarrow & F_2 & \longrightarrow & F_3 & \longrightarrow & 0 \\
& & \cap & & \cap & & \cap & & \\
0 & \longrightarrow & G_1 & \longrightarrow & G_2 & \longrightarrow & G_3 & \longrightarrow & 0
\end{array}
$$

then it may be completed to the commutative exact diagram

$$
\begin{array}{ccccccccc}
& & 0 & & 0 & & 0 & & \\
& & \downarrow & & \downarrow & & \downarrow & & \\
0 & \longrightarrow & F_1 & \longrightarrow & F_2 & \longrightarrow & F_3 & \longrightarrow & 0 \\
& & \downarrow & & \downarrow & & \downarrow & & \\
0 & \longrightarrow & G_1 & \longrightarrow & G_2 & \longrightarrow & G_3 & \longrightarrow & 0 \\
& & \downarrow & & \downarrow & & \downarrow & & \\
0 & \longrightarrow & G_1/F_1 & \longrightarrow & G_2/F_2 & \longrightarrow & G_3/F_3 & \longrightarrow & 0 \\
& & \downarrow & & \downarrow & & \downarrow & & \\
& & 0 & & 0 & & 0 & &
\end{array}
$$

4.4 Direct limits of sheaves

Let $(F_i)_{i \in I}$ be a direct system of abelian sheaves on a topological space X. For any open subset V of X, we have a direct system $(F_i(V))_{i \in I}$

of abelian groups, which is compatible with restriction to smaller open subsets. We may define an abelian presheaf, pre-$\varinjlim F_i$, by letting $(\text{pre-}\varinjlim (F_i)(V) \equiv \varinjlim (F_i(V))$ for each open subset V of X. The restriction res_U^V in this presheaf is given by the $\varinjlim$ (res_U^V in F_i) for each open subset U of V.

We may also define an abelian sheaf, $\varinjlim F_i$ by letting $\varinjlim F_i$ be the associated sheaf, $(\text{pre-}\varinjlim F_i)^s$. It is natural to ask how close $(\text{pre-}\varinjlim F_i)$ is to being a sheaf, but, first, we will study stalks.

Lemma 4.4.1. *Let x be a point of a space X. Let $(F_i)_{i \in I}$ be a direct system of abelian sheaves on X. Then we have natural isomorphisms,*

$$\varinjlim (F_{i,x}) \xrightarrow{\approx} (\text{pre-}\varinjlim (F_i))_x \xrightarrow{\approx} \varinjlim (F_i))_x.$$

Proof. By the definition of stalks, using the fact that two compatible direct limit processes commute, we find the isomorphisms, $\varinjlim (F_{i,x}) = \varinjlim (\varinjlim F_i(V)) \approx \varinjlim (\varinjlim F_i(V)) = \varinjlim ((\text{pre-}\varinjlim F_i)(V)) = (\text{pre-}\varinjlim F_i)_x$. This gives the first isomorphism. The second is a general fact about the sheafification process (see Lemma 4.2.2(a)).   □

Let $(\alpha_i) : (F_i) \to (G_i)$ be a homomorphism between two direct systems of abelian sheaves on X. Then it induces homomorphisms, $\text{pre-}\varinjlim F_i \to \text{pre-}\varinjlim G_i$ and $\varinjlim F_i \to \varinjlim G_i$. Using this concept, we have

Corollary 4.4.2. *Given two homomorphisms $(F_i) \xrightarrow{(\alpha_i)} (G_i) \xrightarrow{(\beta_i)} (H_i)$ of direct systems of abelian sheaves such that, for each i, the sequence $0 \to F_i \xrightarrow{\alpha_i} G_i \xrightarrow{\beta_i} H_i \to 0$ is exact, then the induced sequence,*

$$0 \to \varinjlim F_i \to \varinjlim G_i \to \varinjlim H_i \to 0,$$

is exact.

Proof. To see that this sequence is exact, we may check the exactness of stalks at any point x of X. By hypothesis we have an exact sequence, $0 \to F_{i,x} \to G_{i,x} \to H_{i,x} \to 0$ for each i. By the exactness of direct limits, we have an exact sequence,

$$0 \to \varinjlim (F_{i,x}) \to \varinjlim (G_{i,x}) \to \varinjlim (H_{i,x}) \to 0.$$

Thus, by Lemma 4.4.1, we have an exact sequence,

$$0 \to (\varinjlim F_i)_x \to (\varinjlim G_i)_x \to (\varinjlim H_i)_x \to 0,$$

which is what we wanted.   □

Recall that a topological space X is noetherian if any open subset of X is quasi-compact. For these special topological spaces, we have

Lemma 4.4.3. *Let (F_i) be a direct system of abelian sheaves on a noetherian space X. Then, we have a natural isomorphism, pre-$\varinjlim F_i$ $\approx \varinjlim F_i$.*

Proof. Let V and W be two open subsets of X. For each i, we have an exact sequence,

$$0 \to F_i(V \cup W) \to F_i(V) \oplus F_i(W) \to F_i(V \cap W),$$

as F_i is a sheaf. As direct limits are exact, we have an exact sequence,

$$0 \to \text{pre-}\varinjlim(F_i)(V \cup W) \to \text{pre-}\varinjlim(F_i)(V)\oplus$$

$$\text{pre-}\varinjlim(F_i)(W) \to \text{pre-}\varinjlim(F_i)(V \cap W).$$

Thus, pre-$\varinjlim(F_i)$ satisfies the patching condition for the union of two open subsets of X. By induction, it must satisfy this condition for a finite union of open subsets. As X is noetherian, any union of open subsets is actually a union of a finite number. Therefore, the patching conditions are satisfied in general and, hence, pre-$\varinjlim(F_i)$ is a sheaf equal to its sheafification $\varinjlim F_i$.   □

Corollary 4.4.4. *On a noetherian space, $\varinjlim F_i$ is flabby if each F_i is flabby.*

Proof. Let V be an open subset of X. We have surjection $F_i(X) \to F_i(V)$ for each i. Thus, $\varinjlim(F_i(X)) \to \varinjlim(F_i(V))$ is surjective. By Lemma 4.4.3, this means that $(\varinjlim F_i)(X) \to (\varinjlim F_i)(V)$ is surjective; i.e. $\varinjlim F_i$ is flabby.   □

Let F_i be an abelian sheaf on a topological space X for each i in an index set I. We want to consider $\bigoplus_{i \in I} F_i$ of abelian sheaves on X. By definition $\bigoplus_{i \in I} F_i$ is the sheaf associated to the presheaves $U \to \bigoplus_{i \in I}(F_i(U))$

with coordinatewise restriction which is clearly a decent presheaf. If I is finite, then the presheaf $= \bigoplus_{i \in I} F_i$. If I is infinite they are different in general because a section $\bigoplus_{i \in I} F_i$ is a vector of sections of each F_i which locally has only finitely many non-zero entries but not necessarily globally.

In general $\bigoplus_{i \in I} F_i = \varinjlim_{\substack{S \subset I \\ S\text{finite}}} \bigoplus_{i \in S} F_i$.

A special case of Lemma 4.4.3 is

Corollary 4.4.5. *If X is noetherian then for any open subset U of X,*

$$(\bigoplus_{i \in I} F_i)(U) = \bigoplus_{i \in I} (F_i(U)).$$

Notation. If $\mathcal{F}$ is a sheaf on a topological space X, the *support* of $\mathcal{F}$ is $\{x \in X | \mathcal{F}_x \neq 0\}$. If U is an open subspace of X, $\mathcal{F}|_U$ is the sheaf on U defined by $\mathcal{F}|_U(W) = \mathcal{F}(W)$ if W is open in U with the same restrictions as in $\mathcal{F}$.

5

Sheaves in algebraic geometry

5.1 Sheaves of rings and modules

The most basic concept in algebraic geometry is the concept of a sheaf of rings. If X is a topological space, a *sheaf of rings* $\mathcal{A}$ on X is a sheaf such that for each open subset U of X, $\mathcal{A}(U)$ is a ring and restrictions are ring homomorphisms. Thus the stalk $\mathcal{A}_x$ at each point x of X is naturally a ring.

If X is a space with functions the structure sheaf $\mathcal{O}_X$ is a sheaf of k-algebras (in particular rings). If x is a point of X, the k-algebra $\mathcal{O}_{X,x}$ is a local ring with maximal ideal m_x consisting of germs of functions which vanish at x. This is a local ring because any element of $\mathcal{O}_{X,x} - m_x$ is a unit.

A secondary concept is that of *a sheaf of modules*. Let $\mathcal{A}$ be a sheaf of rings on a topological space X. A sheaf of $\mathcal{A}$-modules is an abelian sheaf $\mathcal{M}$ such that for any open subset U of X the set $\mathcal{M}(U)$ is an $\mathcal{A}(U)$-module and restriction respects multiplication; i.e. $(a \cdot m)|_V = a|_V \cdot m|_V$ for open V in U. The notion of homomorphism of $\mathcal{A}$-modules is the obvious one.

The simplest kind of $\mathcal{A}$-modules are $\mathcal{A}^{\oplus I}$ the direct sum of $\mathcal{A}$ with itself I times with multiplication $\alpha(\beta_i)_{i \in I} = (\alpha\beta_i)_{i \in I}$ and vector addition as usual where we require that the vectors (β_i) locally have only a finite number of non-zero coefficients β_i. These are called *free $\mathcal{A}$-modules*. If I is finite, then $\# I$ is called the *rank* of $\mathcal{A}^{\oplus I}$.

More generally an $\mathcal{A}$-module $\mathcal{M}$ is *locally free* if there is an open cover $X = \bigcup X_i$ such that $\mathcal{M}|_{X_i}$ is a free $\mathcal{A}|_{X_i}$-module. Thus furthermore a locally free $\mathcal{A}$-module is said to be *of finite rank n* if all the $\mathcal{M}|_{X_i}$ have rank n. A locally free $\mathcal{A}$-module of rank one is said to be *invertible*.

An $\mathcal{A}$-module $\mathcal{M}$ is said to be *quasi-coherent* if it is locally given by generators and relations; i.e. there is an open cover $X = \bigcup X_i$ such that we have an exact sequence of $\mathcal{A}|_{X_i}$-modules

$$\mathcal{A}|_{X_i}^{\oplus J} \overset{\psi_i}{\to} \mathcal{A}|_{X_i}^{\oplus I} \to \mathcal{M}|_{X_i} \to 0.$$

Here ψ_i is given by a $I \times J$ matrix of sections of $\mathcal{A}$ over X_i where the rows with fixed j locally only have a finite number of non-zero entries. Thus if X is noetherian the matrix ψ_i is given by an $\mathcal{A}(X_i)$-linear transformation $\psi_i : \mathcal{A}(X_i)^{\oplus J} \to \mathcal{A}(X_i)^{\oplus I}$.

Let M be an $\mathcal{A}(X)$-module. Then we can form an $\mathcal{A}$-module $M \otimes_{\mathcal{A}(X)} \mathcal{A}$ by taking it to be the sheaf associated to the presheaf

$$U \to M \otimes_{\mathcal{A}(X)} \mathcal{A}(U).$$

Clearly if $M_1 \to M_2 \to M_3 \to 0$ is an exact sequence of $\mathcal{A}(X)$-modules, the sequence

$$M_1 \otimes_{\mathcal{A}(X)} \mathcal{A} \to M_2 \otimes_{\mathcal{A}(X)} \mathcal{A} \to M_3 \otimes_{\mathcal{A}(X)} \mathcal{A} \to 0$$

is exact. In particular if M is the cokernel of a homomorphism

$$\psi : \mathcal{A}(X)^{\oplus J} \to \mathcal{A}(X)^{\oplus I},$$

then we have an exact sequence

$$\mathcal{A}^{\oplus J} \overset{\psi}{\to} \mathcal{A}^{\oplus I} \to M \otimes_{\mathcal{A}(X)} \mathcal{A} \to 0.$$

Thus on a noetherian space $\mathcal{M}$ is quasi-coherent iff it locally has the form $M_i \otimes_{\mathcal{A}(X_i)} (\mathcal{A}|_{X_i})$.

Exercise 5.1.1. Let $\mathcal{M}$ be a sheaf of $\mathcal{A}$-modules. For each n define $\mathrm{Sym}^n \mathcal{M}$ to be the sheaf associated to the presheaf $U \to \mathrm{Sym}^n{}_{\mathcal{A}(U)} \mathcal{M}(U)$. Show that $\mathrm{Sym}^n \mathcal{M}$ is locally free if $\mathcal{M}$ is locally free. Define the exterior power $\Lambda^n \mathcal{M}$ and show the same.

Exercise 5.1.2. Let $\mathcal{N}$ and $\mathcal{M}$ be sheaves of $\mathcal{A}$-modules. Define sheaves $\mathrm{Hom}_{\mathcal{A}}(\mathcal{N}, \mathcal{M})$ and $\mathcal{N} \otimes_{\mathcal{A}} \mathcal{M}$.

Exercise 5.1.3. Let $\mathcal{M}$ be a locally free $\mathcal{A}$-module of constant rank r. Then $\det \mathcal{M} = \Lambda^r \mathcal{M}$ is an invertible $\mathcal{A}$-module. If $0 \to \mathcal{M}_1 \to$

$\mathcal{M}_2 \to \mathcal{M}_3 \to 0$ is an exact sequence of such $\mathcal{A}$-modules, then we have a canonical isomorphism

$$\det \mathcal{M}_2 \approx \det \mathcal{M}_1 \otimes_{\mathcal{A}} \det \mathcal{M}_3.$$

5.2 Quasi-coherent sheaves on affine varieties

Let X be an affine variety with the structure sheaf $\mathcal{O}_X$.

We have the following true statements:

(1) $k[X] \overset{\approx}{\to} \Gamma(X, \mathcal{O}_X)$,

(2)   for any f in $k[X]$, we have $k[X]_{(f)} \overset{\approx}{\to} \Gamma(D(f), \mathcal{O}_X)$,

(3)   for any x in X we have $k[X]_{n_x} \overset{\approx}{\to} \mathcal{O}_{X,x}$ where n_x is the maximal ideal of functions in $k[X]$ vanishing at x.

Here (3) follows from (2) by taking limits over smaller and smaller affine neighborhoods $D(f)$ of x.

A next objective is to generalize the above results to a whole class of sheaves. Let M be a $k[X]$-module. Then the sheaf $M \otimes_{k[X]} \mathcal{O}_X$ is denoted by $\tilde{M}$. Then we have

Proposition 5.2.1.

(a) $M \overset{\approx}{\to} \Gamma(X, \tilde{M})$,

(b)   *for any f in $k[X]$, $M_{(f)} \overset{\approx}{\to} \Gamma(D(f), \tilde{M})$, and*

(c)   *for any x in X, $M_{n_x} \overset{\approx}{\to} (\tilde{M})_x$.*

Proof. For (c), $\tilde{M}_x = M \otimes_{k[X]} \mathcal{O}_{X,x} = M \otimes_{k[X]} k[X]_{n_x} = M_{n_x}$. Now (b) is a special case of (a) because

$$\tilde{M}|_{D(f)} = M \otimes_{k[X]} \mathcal{O}_X|_{D(f)} = M_{(f)} \otimes_{k[X]_{(f)}} \mathcal{O}_{D(f)}.$$

We will first show that $M \to \Gamma(X, \tilde{M})$ is injective and then use the injectivity in (b) to show that $M \to \Gamma(X, \tilde{M})$ is surjective.

For the injectivity let m be in the kernel. By (c) this means for all x in X we have a function f_x in $k[X]$ which does not vanish at x such that $f_x m = 0$. Thus the $D(f_x)$ covers X. So by the nullstellensatz find a_i in $k[X]$ such that $1 = \sum a_i f_{x_i}$. Thus $m = 1 \times m = \sum a_i f_{x_i} m = \sum 0 = 0$. This shows injectivity.

To show surjectivity let $X = \bigcup D(f_i)$ be an open cover of X. A section of $\tilde{M}$ over X is gotten by pasting together a section α_i over $D(f_i)$ from $M_{(f_i)}$ such that $\alpha_i = \alpha_j$ in $M_{(f_i \cdot f_j)}$. Replacing f_i by a power we may assume that $\alpha_i = \frac{m_i}{f_i}$ with m_i in M. The patching conditions means $(f_i m_j - f_j m_i)(f_i f_j)^N = 0$. As we may assume that the cover is finite we may have a uniform N for all i and j. Replace f_i by f_i^{N+1} and m_i by $f_i^N m_i$. Then we have $f_i m_j = f_j m_i$. Find a_i in $k[X]$ such that $1 = \sum a_i f_i$. Take $m = \sum a_j m_j$. We have $f_i m = f_i \sum a_j m_j = \sum a_j f_i m_j = (\sum a_i f_i) m_i = 1 \times m_i = m_i$. Thus $m = m_i/f_i$ locally. Thus $M \to \Gamma(X, \tilde{M})$ is surjective. $\qquad\square$

We have seen in the last section that $\tilde{M}$ is quasi-coherent. We want to prove the converse. This will use a direct sequence of sheaves with studies of the "poles" on X of sections of sheaves over $D(f)$.

Let U be an open subset of a topological space X. Let $\mathcal{F}$ be a sheaf on X. Define a new sheaf $_U\mathcal{F}$ on X by the rule $_U\mathcal{F}(V) = \mathcal{F}(U \cap V)$ with the obvious restrictions. Restriction to U defines a sheaf mapping $\mathcal{F} \to {}_U\mathcal{F}$ over X.

Let X be a variety. Let f be a regular function on X. Let $\mathcal{F}$ be an $\mathcal{O}_X$-module. Let $\frac{1}{f^i}$ be a bookkeeping symbol. We have a sequence $\mathcal{F} = \frac{1}{1}\mathcal{F} \to \frac{1}{f}\mathcal{F} \to \frac{1}{f^2}\mathcal{F} \to \ldots \to$ where arrows send $\frac{1}{f^i}(\alpha)$ to $\frac{1}{f^{i+1}}(f\alpha)$. The direct limit of this sequence will be denoted by $\mathcal{F}_{(f)}$. We have a canonical homomorphism $\mathcal{F}_{(f)} \to_{D(f)} \mathcal{F}$ which sends $\frac{1}{f^i}(\alpha) = \left(\frac{1}{f|_{D(f)}}\right)^i (\alpha|_{D(f)})$.

Proposition 5.2.2.

(a)   An $\mathcal{O}_X$-module $\mathcal{F}$ on an affine variety X is quasi-coherent

iff

(b)   for all f in $k[X]$ the homomorphism $\mathcal{F}_{(f)} \to_{D(f)} \mathcal{F}$ is an isomorphism.

iff

(c) $\mathcal{F} \approx \tilde{M}$ for some $k[X]$-module M.

Proof. $(c) \Rightarrow (a)$ is trivial. We will show that $(a) \Rightarrow (b)$ and $(c) \Leftrightarrow (b)$.

Assume that (a) is true. Then $\mathcal{F}$ locally has the form $\tilde{M}$ and the statement (b) is local on X. Thus we may assume that $\mathcal{F} = \tilde{M}$ where M is an $k[X]$-module. We want to show that $(c) \Rightarrow (b)$. The key fact is

Lemma 5.2.3. *If U is an open subset of X, then*
$$\Gamma(U, \mathcal{F}_{(f)}) = \Gamma(U, \mathcal{F})_{(f)}.$$

Proof. By Lemma 4.4.3, $\Gamma(U, \mathcal{F}_{(f)}) = \varinjlim_{n \to \infty} \frac{1}{f^n}\Gamma(U, \mathcal{F}) = \Gamma(U, \mathcal{F})_{(f)}$.

$\qquad\square$

Now let $D(g)$ be a small open subset of X.

$\Gamma(D(g), \tilde{M}_{(f)}) = \Gamma(D(g), \tilde{M})_{(f)} = (M_{(g)})_{(f)} = M_{(gf)} = \Gamma(D(gf), \tilde{M})$. This proves (b). Conversely to show that (b)$\Rightarrow$(c). Let $M = \Gamma(X, \mathcal{F})$. We have a canonical $\mathcal{O}_X$-homomorphism $\psi : \tilde{M} \to \mathcal{F}$, which we want to show is an isomorphism. Now $\Gamma(D(g), \tilde{M}) = M_{(g)}$ but $\Gamma(D(g), \mathcal{F}) = \Gamma(X, _{D(g)}\mathcal{F}) = \Gamma(X, \mathcal{F}_{(g)}) = \Gamma(X, \mathcal{F})_{(g)} = M_{(g)}$. Thus ψ is an isomorphism. $\qquad \square$

Thus gives a functor $\sim$ from $\{k[X]\text{-modules}\}$ to $\{$quasi-coherent sheaves $\mathcal{F}\}$. The inverse of $\sim$ is $\Gamma(X, \ )$. Thus we have an equivalence of categories. As for exactness

Proposition 5.2.4. *On an affine variety X*

(a)   *if $0 \to M_1 \to M_2 \to M_3 \to 0$ is an exact sequence of $k[X]$-modules then $0 \to \tilde{M}_1 \to \tilde{M}_2 \to \tilde{M}_3 \to 0$ is an exact sequence of quasi-coherent $\mathcal{O}_X$-modules, and*

(b)   *if $0 \to \mathcal{F}_1 \to \mathcal{F}_2 \to \mathcal{F}_3 \to 0$ is an exact sequence of quasi-coherent $\mathcal{O}_X$-modules then $0 \to \Gamma(X, \mathcal{F}_1) \to \Gamma(X, \mathcal{F}_2) \to \Gamma(X, \mathcal{F}_3) \to 0$ is an exact sequence of $k[X]$-modules.*

Proof. For (a) just check the exactness of stalks at a point x. We want $0 \to M_{1,n_x} \to M_{2,n_x} \to M_{3,n_x} \to 0$ to be exact but localization is exact. Thus (a) is true.

For (b) in general we have an exact sequence of $k[X]$-modules,

$$0 \to \Gamma(X, \mathcal{F}_1) \to \Gamma(X, \mathcal{F}_2) \to \Gamma(X, \mathcal{F}_3) \to M \to 0.$$

Applying $\tilde{\ }$ we see that $\tilde{M} = 0$. Hence $M = \Gamma(X, 0) = 0$. This shows that (b) is true. $\qquad \square$

Corollary 5.2.5. *Let X be any variety. Let $\psi : \mathcal{F} \to \mathcal{G}$ be an $\mathcal{O}_X$-homomorphism between two quasi-coherent $\mathcal{O}_X$-modules. Then $\mathrm{Ker}(\psi)$ and $\mathrm{Cok}(\psi)$ are quasi-coherent $\mathcal{O}_X$-modules.*

Proof. The statement is local on X so we may assume that X is affine. Then by the above $\mathrm{Ker}(\psi) = \widetilde{\mathrm{Ker}(\Gamma(X, \psi))}$ and $\mathrm{Cok}(\psi) = \widetilde{\mathrm{Cok}(\Gamma(X, \psi))}$. $\qquad \square$

5.3   Coherent sheaves

Let $\mathcal{A}$ be a sheaf of rings on a topological space X. An $\mathcal{A}$-module $\mathcal{M}$ is *coherent* if locally it has a presentation $\mathcal{A}|_U^{\oplus I} \to \mathcal{A}|_U^{\oplus J} \to \mathcal{M}|_U \to 0$ where I and J are finite sets.

Now if X is a variety an $\mathcal{O}_X$-module $\mathcal{M}$ is coherent iff $\mathcal{M}$ locally has the form $\tilde{M}$ where M is a finitely generated $k[U]$-module where U is an open affine subvariety of X. This equivalence uses the fact that $k[U]$ is a noetherian ring.

Next we have another local to global result.

Lemma 5.3.1. *Let M be a $k[X]$-module where X is an affine variety. Then $\tilde{M}$ is coherent if and only if M is a finitely generated $k[X]$-module.*

Proof. The "if" part is clear. For the converse we may assume that we have a finite open cover $X = \bigcup D(f_i)$ where $\tilde{M}|_{D(f_i)} = \tilde{N}_i$ where N_i is a $k[X]_{(f_i)}$-module of finite type. Here $N_i = \Gamma(D(f), \tilde{M}) = M_{(f_i)}$. Thus for each i we have a finite number of $M_{i,j} = \frac{m_{i,j}}{f_i^{j,j}}$ where $m_{i,j}$ are in M which span the $k[X]_{(f_i)}$-module $M_{(f_i)}$. Let M_1 be the $k[X]$-sub-module generated by the finitely many $m_{i,j}$. Now $\alpha : \tilde{M}_1 \hookrightarrow \tilde{M}$ but α is locally surjective by construction. Hence α and consequently $\Gamma(X, \alpha) : M_1 \to M$ is an isomorphism. $\square$

Exercise 5.3.2. Prove that a quasi-coherent $\mathcal{O}_X$-module on a variety X which is contained in a coherent one is also coherent.

Frequently if $i : X \subset Y$ is a closed subvariety of a variety, we identify an $\mathcal{O}_X$-module $\mathcal{F}$ on X with an $\mathcal{O}_Y$-module $\mathcal{F}'$ on Y. By definition $\mathcal{F}'(-) = \mathcal{F}(X \cap -)$ with the obvious restrictions and multiplication $f \cdot \alpha = (i^* f) \cdot \alpha$. The stalks of $\mathcal{F}'$ are zero at points in the complement of X and $\mathcal{F}_x = \mathcal{F}'_x$ at x in X. Thus $\mathcal{F}'$ is said to be *supported by* the set X. $\mathcal{F}'$ is a (quasi-)coherent $\mathcal{O}_Y$-module if and only if $\mathcal{F}$ is a (quasi-)coherent $\mathcal{O}_X$-module. This can be checked locally when X and Y are affine. In this case $(\tilde{M})' = (\widetilde{M'})$ for each $k[X]$-module M where M' is the same group on which $k[Y]$ acts via i^*. We will drop the prime in practice.

We have a k-algebra homomorphism of sheaves on Y : $\mathcal{O}_Y \to \mathcal{O}_X$ which is surjective. Let $\mathcal{I}_X$ be the kernel; then $\mathcal{I}_X$ is the *ideal sheaf* of regular functions on Y which vanish on X. We thus have an exact sequence

$$0 \to \mathcal{I}_X \subset \mathcal{O}_Y \to \mathcal{O}_X \to 0$$

as $\mathcal{I}_X$ is a coherent $\mathcal{O}_Y$-module because it is quasi-coherent and contained in $\mathcal{O}_Y$. We can characterize the $\mathcal{O}_Y$-modules of the form $\mathcal{F}'$ as those $\mathcal{O}_Y$-modules on which $\mathcal{I}_X$ acts by zero. Let $\mathcal{G}$ be any quasi-coherent $\mathcal{O}_Y$-module. Then the quotient $\mathcal{G}/\mathcal{I}_X\mathcal{G} = \mathcal{G}|_X$ always has this form.

Let X be a variety. Consider the diagonal embedding $X \hookrightarrow X \times X$.

We have an open subset U of $X \times X$ such that X is closed in U. Let Ω_X be the coherent sheaf of $\mathcal{O}_X$-modules corresponding to $\mathcal{I}_X/\mathcal{I}_X^2 = \mathcal{I}_X|_\Delta$. This is clearly independent of the choice of U. The sheaf Ω_X is called the *sheaf of differentials* which we will study in detail later.

There are some interesting invertible sheaves on the projective space $\mathbf{P}^n$. Let $\pi : \mathbf{A}^{n+1} - \{0\} \to \mathbf{P}^n$ be the projection. For any integer m let $\mathcal{O}_{\mathbf{P}^n}(m)(U) = \{$regular functions on $\pi^{-1}U$ which are homogeneous of degree $m\}$. Thus $\mathcal{O}_{\mathbf{P}^n} = \mathcal{O}_{\mathbf{P}^n}(0)$ and we have natural homomorphisms given by multiplications $\mathcal{O}_{\mathbf{P}^n}(m_1) \otimes_{\mathcal{O}_{\mathbf{P}^n}} \mathcal{O}_{\mathbf{P}^n}(m_2) \to \mathcal{O}_{\mathbf{P}^n}(m_1+m_2)$ which are isomorphisms. Now $\mathcal{O}_{\mathbf{P}^n}(m)|_{\{X_i \neq 0\}} = X_i^m \cdot \mathcal{O}_{\mathbf{P}^n}$. Thus $\mathcal{O}_{\mathbf{P}^n}(m)$ are invertible and calculations are easy. If X is a closed subvariety of $\mathbf{P}^n, \mathcal{O}_{\mathbf{P}^n}(m)|_X$ is denoted by $\mathcal{O}_X(m)$. More generally if $\mathcal{F}$ is an $\mathcal{O}_X$-module $\mathcal{F}(m) = \mathcal{F} \otimes_{\mathcal{O}_X} \mathcal{O}_X(m)$. We shall later see that any invertible sheaf on $\mathbf{P}^n$ is isomorphic to $\mathcal{O}_{\mathbf{P}^n}(m)$ where m is uniquely determined if $n > 0$.

Let x be a point of a variety X and $\mathcal{F}$ be a coherent sheaf on X. Let $\mathcal{F}|_x$ be the vector space $\mathcal{F}_x/m_x\mathcal{F}_x$ at the point x.

If σ is a section of $\mathcal{F}$ over a neighborhood we denote the image of σ_x in $\mathcal{F}|_x$ by $\sigma(x)$. A useful form of Nakayama's lemma is

Lemma 5.3.3. *If $\mathcal{F}$ is a coherent sheaf on a variety X, let x be a point of X. Then $\mathcal{F}|_U = 0$ for some neighborhood U of x if and only if $\mathcal{F}|_x = 0$.*

Proof. The "only if" part is clear. Conversely we may assume that X is affine and $\mathcal{F} = \tilde{M}$ where M is a finitely generated $k[X]$-module. Let n_x be the ideal of $k[X]$ of functions vanishing at x.

Claim. $\mathcal{F}|_x = M/n_x M$.

As $M/n_x M$ is a $k[X]$-module on which $k[X] - n_x$ acts invertibly, $(M/n_x M)_{n_x} = M/n_x M$ but the first $M_{n_x}/n_{x_{n_x}} M_{n_x} = \mathcal{F}_x/m_x\mathcal{F}_x$. This proves the claim. Thus we have $M = n_x M$.

By Nakayama's Lemma 1.4.3, there is a regular functions f such that $f(x) \neq 0$ and $fM = 0$. Thus $M_{(f)} = 0$. Hence $U = D(f)$ is a neighborhood of x such that $\mathcal{F}|_U = \tilde{M}_{(f)} = 0$. $\qquad\square$

Corollary 5.3.4. *In the situation of the lemma*

(a) *Let $\sigma_1, \ldots, \sigma_n$ be sections of $\mathcal{F}$. Then the homomorphism $\psi : \mathcal{O}_X^{\oplus n} \to \mathcal{F}$ sending $e_i \to \sigma_i$ is surjective in a neighborhood of x if and only if $\sigma_1(x), \ldots, \sigma_n(x)$ span $\mathcal{F}|_x$.*

(b)   The function $x \to \dim_k(\mathcal{F}|_x)$ is *upper-semicontinuous; i.e. the subsets* $\{x \in X \mid \dim_k(\mathcal{F}|_x) \geq m\}$ *are closed.*

(c)   *This function has constant value m if and only if $\mathcal{F}$ is locally free of rank m.*

Proof. (a) is the lemma applied to Cok ψ as Cok $\psi|_x = \mathcal{F}|_x / \sum k\sigma_i(x)$. For (b) let $n = \dim \mathcal{F}|_x$. We want to show that the set $\{y \in X \mid \dim \mathcal{F}|_y \leq n\}$ contains a neighborhood of x. Choose a basis $\sigma_1(x), \ldots, \sigma_n(x)$ which span $\mathcal{F}|_x$ where $\sigma_1, \ldots, \sigma_n$ are sections over a neighborhood of x. Then by (a) $\sigma_1(y), \ldots, \sigma_n(y)$ span $\mathcal{F}|_y$ for y in a neighborhood of x. Thus (b) is true. For (c) the "if" part is clear. For the "only if" let $\sigma_1, \ldots, \sigma_m$ be local sections near a point x such that $\sigma_1(x), \ldots, \sigma_m(x)$ are a basis of $\mathcal{F}|_x$. Then we have a surjection $\psi : \mathcal{O}_U^{\oplus m} \to \mathcal{F}|_u \to 0$ in a neighborhood U of x. By our dimension assumption $\psi|_y$ is an isomorphism for all y in U. Let $(f_1, \ldots, f_m)$ be a section of Ker(ψ). Then $f_i(y) = 0$ for all i and y in U. Thus $f_i = 0$, hence Ker $\psi = 0$ and ψ is an isomorphism. $\square$

Remark. The last fact is not true of schemes in general.

5.4   Quasi-coherent sheaves on projective varieties

Let X be a projective variety in $\mathbf{P}^n$ and $C(X)$ be the cone over X. Then we have the projection $C(X) - \{0\} \to X$. The ring $k[C(X)]$ is graded. Let M be a graded $k[C(X)]$-module. We want to define a sheaf $\tilde{M}$ on X. We have the old sheaf $\tilde{M}$ on $C(X)$. Let $\tilde{\tilde{M}} = \tilde{M}|_{C(X)-0}$. Let U be an open subset of X. Then $\tilde{M}(\pi^{-1}U)$ is graded. By definition $\tilde{M}(U) = (\tilde{\tilde{M}}(\pi^{-1}U))_{\text{degree } 0}$. Explicitly if f is a homogeneous element of $k[C(X)]$ then $\tilde{M}|_{D(f)} = \widetilde{M_{(f) \text{ degree } 0}}$ where $D(f) =$ Spec $k[C(X)]_{(f) \text{ degree } 0}$. Thus $\tilde{M}$ is quasi-coherent on X and it is coherent if M is finitely generated. We intend to show

Theorem 5.4.1. *All quasi-coherent sheaves on X have the form $\tilde{M}$. If the sheaf is coherent it has the form $\tilde{M}$ where M is a finitely generated graded $k[C(X)]$-module.*

Remark. One must be careful because $\tilde{M}$ does not determine M.

Exercise 5.4.2. Give an example of a non-zero M such that $\tilde{M} = 0$.

Proof.   Given a quasi-coherent sheaf $\mathcal{F}$ on X, we want to construct a graded module. We have a graded ring homomorphism

$$\psi : k[C(X)] \to \bigoplus_{n \geq 0} \Gamma(X, \mathcal{O}_X(n))$$

. Thus it will suffice to construct a module over this ring. Consider $M = \bigoplus_{n \geq 0} \Gamma(X, \mathcal{F}(n))$.

Claim.   We have a natural isomorphism $\psi : \tilde{M} \overset{\cong}{\to} \mathcal{F}$.

Let us see what the map is locally. Let f be a degree d homogeneous regular function on $C(X)$. $\tilde{M}|_{D(f)} = M_{(f) \text{ degree } 0}$. Here an element of $M_{(f) \text{ degree } 0}$ is given by a fraction $\frac{\alpha}{f^i}$ where α is a section $\Gamma(X, \mathcal{F}(id))$. This defines a section of $\mathcal{F}$ over $D(f_i)$. This defines ψ locally as it is clearly compatible with localization. More suggestively

$$M_{(f) \text{ degree } 0} = \varinjlim \left(\Gamma(X, \mathcal{F}) \overset{f}{\to} \frac{1}{f} \Gamma(\mathcal{F}(d)) \to \dots \frac{1}{f^i} \Gamma(\mathcal{F}(id) \to \dots) \right)$$

$$= \Gamma(X, \varinjlim (\mathcal{F} \to \frac{1}{f} \mathcal{F}(d) \to \dots)) \text{ [Lemma 4.4.3]}$$

$$= \Gamma(X, _{D(f)} \mathcal{F}) \text{ [Proposition 5.2.2]} = \Gamma(D(f), \mathcal{F}).$$

This shows that the mapping is an isomorphism.

If $\mathcal{F} = \tilde{M}$ is coherent then $\mathcal{F} = \bigcup \tilde{M}_i$ where M_i ranges through the finitely generated graded $k[C(X)]$-submodules of M. As $\tilde{M}_i + \tilde{M}_j = (M_i + M_j)\tilde{\ }$ and $\mathcal{F}$ is coherent, it follows that $\mathcal{F} = \tilde{M}_i$ for some i by the ascending chain condition as X is quasi-compact. $\square$

Corollary 5.4.3.   *If $\mathcal{F}$ is a coherent sheaf on a projective variety $X \subset \mathbf{P}^n$ then there exists n_0 such that we have a surjection $\mathcal{O}_X^{\oplus \text{finite}} \to \mathcal{F}(n) \to 0$ when $n \geq n_0$.*

Proof.   Just take $n_0 = $ max degree of generators of M in the theorem. $\square$

5.5   Invertible sheaves

Let X be a variety. By definition Pic X is the *group of isomorphism classes of invertible sheaves* on X with tensor product as group law. This group is an important invariant of a variety. In this section we want to develop some methods of computing Pic X. We will assume that X is irreducible.

Let $k(X) = \bigcup_{\substack{\emptyset \neq U \subset X \\ \text{open}}} k[U]$ be the field of rational functions on X. We

define a sheaf of rings $\mathbf{Rat}_X$ on X by the rule that for any non-empty subset U of X, $\mathbf{Rat}_X(U) = k(X)$ with trivial restrictions. Then $\mathcal{O}_X$ is a subsheaf of $\mathbf{Rat}_X$. If X is affine then $\mathbf{Rat}_X = \widetilde{k(X)}$. Thus $\mathbf{Rat}_X$ is a quasi-coherent $\mathcal{O}_X$-module in general.

A *sheaf of fractional ideals* $\mathcal{I}$ is a coherent subsheaf $\mathcal{I} \subset \mathbf{Rat}_X$. Let IFI$(X)$ denote the group under multiplication of all *invertible fractional ideals*. A fractional ideal is *principal* if it has the form $f \cdot \mathcal{O}_X$ where f is a rational function on X. Let $P(X)$ be the *set* of *principal ideals*. Then $P(X)$ is a subgroup of IFI(X). We have a homomorphism $\psi : \text{IFI}(X) \rightarrow \text{Pic}(X)$ which sends an invertible fractional ideal into its isomorphism class. Clearly ψ is a homomorphism.

Lemma 5.5.1. *We have an exact sequence*
$$1 \rightarrow P(X) \rightarrow \text{IFI}(X) \overset{\psi}{\rightarrow} \text{Pic}(X) \rightarrow 1.$$

Proof. The kernel of ψ consists of invertible fractional ideals $\mathcal{I}$ such that $\mathcal{O}_X \approx \mathcal{I}$. Let f be the image of 1 under this isomorphism. Then $\mathcal{I} = f \cdot \mathcal{O}_X$; i.e., $\mathcal{I}$ is principal. Conversely if $\mathcal{I}$ is principal $\mathcal{I} \approx \mathcal{O}_X$. Thus $P(X) = \text{Ker}(\psi)$.

It remains to prove that ψ is surjective. So we are given an invertible sheaf $\mathcal{L}$ on X and we want to construct an invertible fractional ideal $\mathcal{I}$ such that $\mathcal{I} \approx \mathcal{L}$. Let σ be a non-zero section of $\mathcal{L}$ over an open dense subset. Let $\mathcal{I}$ be the subsheaf of $\mathbf{Rat}_X$ defined as follows: If U is an open dense subset on X, $\mathcal{I}(U) = \{f \in k(X) | f \cdot \sigma$ comes from a section of $\mathcal{L}$ over $U\}$. It is trivial to check that multiplication by σ defines an isomorphism $\mathcal{I} \overset{\approx}{\rightarrow} \mathcal{L}$. Thus $\mathcal{I}$ is an invertible fractional ideal. $\square$

An *irreducible divisor* D on X is a closed irreducible subvariety $D \subset X$ such that dim $D = $ dim $X - 1$. The group Div(X) is the *free abelian* (additive) group generated by the set of all irreducible divisors on X. A (Weil) *divisor* D is an element of Div(X). Thus $D = \sum_{\text{finite}} n_i D_i$ where coefficients n_i are integers and each D_i is an irreducible divisor. A divisor D is *effective* if all the coefficients $n_i \geq 0$.

Proposition 5.5.2. *Assume that the local rings $\mathcal{O}_{X,x}$ of X at all points x are unique factorization domains. Then we may associate to each divisor D an invertible sheaf $\mathcal{O}_X(D)$ of fractional ideals such that $\mathcal{O}_X(-)$ defines an isomorphism $\text{Div}(X) \overset{\approx}{\rightarrow} \text{IFI}(X)$ and 1 is a global section of $\mathcal{O}_X(D)$ iff D is effective.*

Proof. Let $D = \sum n_i D_i$ where the D_i's are irreducible. Then $\mathcal{O}_X(D)$ must be $\Pi \mathcal{O}_X(D_i)^{n_i}$. Thus to define $\mathcal{O}_X(D)$ we must define $\mathcal{O}_X(D)$ when D is irreducible.

Claim. Let $\mathcal{I}_D$ be the ideal of an irreducible divisor D in X. Then $\mathcal{I}_D$ is invertible.

If we prove this claim we may define $\mathcal{O}_X(D) = \mathcal{I}_D^{-1}$. (Thus $\mathcal{I}_D = \mathcal{O}_X(-D)$).

To prove the claim let f be an irreducible element of $\mathcal{I}_{D,x}$. Taking X to be smaller we may assume that f extends to a regular function $\tilde{f}$ on X and X is affine. Let $(\tilde{f} = 0) = D \cup C$ where C is a union of the other components. We may find a regular function $\tilde{g}$ on X such that $\tilde{g}$ vanishes on C but $\tilde{g}$ does not vanish on D. Let $\tilde{h}$ be any regular function which vanishes on D. Then $\tilde{h} \cdot \tilde{g}$ vanishes on $(\tilde{f} = 0)$. Thus $\tilde{f}$ divides $(\tilde{h} \cdot \tilde{g})^n$ for $n > 0$. Now let us take germs at x by erasing $\sim$. Then $f | h \cdot g$ but $f \backslash g$ as g does not vanish on D. Thus $f | h$ as $\mathcal{O}_{X,x}$ is a unique factorization domain. This proves that $(f) = \mathcal{I}_{D,x}$. So $(\tilde{f})$ is a basis of $\mathcal{I}_D$ in a neighborhood of x. This proves the claim.

Next let $\mathcal{I}$ be an invertible ideal; i.e., $\mathcal{I} \subseteq \mathcal{O}_X$. We want to show

Claim. $\mathcal{I} = \mathcal{O}_X(-D)$ for some effective divisor D.

To prove this, first note that $\mathcal{O}_X = \mathcal{O}_X(0)$. By induction we may assume that the claim is true for all strictly larger invertible ideals and $\mathcal{I} \subsetneq \mathcal{O}_X$. Consider zeroes$(\mathcal{I}) \equiv$ support$(\mathcal{O}_X / \mathcal{I})$. This is a non-empty closed subset of X. By the principal ideal theorem any component of zeroes$(\mathcal{I})$ has dimension $= \dim X - 1$; i.e., it is an irreducible divisor. Let E be one of these components. Then $\mathcal{I}' = \mathcal{I} \cdot \mathcal{O}_X(E)$ is an invertible ideal strictly larger than $\mathcal{I}$. Thus $\mathcal{I}' = \mathcal{O}_X(-D')$ where D' is effective. Hence $\mathcal{I} = \mathcal{O}_X(-D' - E)$. This proves the claim.

Now let $\mathcal{I}$ be an invertible fractional ideal. Then $\mathcal{J} = \mathcal{I} \cap \mathcal{O}_X$ is invertible because each $\mathcal{O}_{X,x}$ is a UFD (a local calculation). Now $\mathcal{J}$ and $\mathcal{J}\mathcal{I}^{-1}$ are invertible ideals and $\mathcal{I} = \mathcal{J} \cdot (\mathcal{J}\mathcal{I}^{-1})^{-1}$. Thus $\mathcal{I}$ has the form $\mathcal{O}_X(D)$. I will leave the last statement as an exercise. $\qquad\square$

Thus we have reduced the problem of determining IFI(X) to the geometry of X when X is locally factorial. We will use this to compute some Pic(X).

Lemma 5.5.3. Pic$(\mathbf{A}^n) = \{\mathcal{O}_{\mathbf{A}_n}\}$.

Proof. As polynomial rings are UFDs all the local rings of $\mathbf{A}^n$ are

UFDs. Thus we may apply Proposition 5.5.2. Thus it suffices to prove that $\mathcal{I}_D$ is principal where D is an irreducible divisor on $\mathbf{A}^n$. Now $D = $ zeroes(f) where f is an irreducible polynomial. Thus $\mathcal{I}_D = f \cdot \mathcal{O}_{\mathbf{A}^n}$ because f generates a prime ideal. $\square$

Next we have

Lemma 5.5.4. *If $n > 0$, Pic($\mathbf{P}^n$) $= \{\mathcal{O}_{\mathbf{P}^n}(m)|m \in \mathbf{Z}\}$ and these sheaves are non-isomorphic; i.e., Pic($\mathbf{P}^n$) is the free abelian group generated by the isomorphism class of $\mathcal{O}_{\mathbf{P}^n}(1)$.*

Proof. Let E be an irreducible divisor on $\mathbf{P}^n$. Then $E = (f = 0)$ where f is an irreducible homogeneous polynomial of degree e. Then by a local calculation $\mathcal{I}_E = $ image of $\mathcal{O}_{\mathbf{P}^n}(-e)$ in $\mathcal{O}_{\mathbf{P}^n}$ under multiplication by f. Hence $\mathcal{I}_E \approx \mathcal{O}_{\mathbf{P}^n}(-e)$. Thus as $\mathcal{I}_E$ generates Pic($\mathbf{P}^n$) any invertible sheaf is isomorphic to a product and hence one of the $\mathcal{O}_{\mathbf{P}^n}(m)$. Now $\Gamma(\mathbf{P}^n, \mathcal{O}_{\mathbf{P}^n}(m))$ is one dimensional if and only if $m = 0$. Thus if $m \neq 0$ then $\mathcal{O}_{\mathbf{P}^n}(p + m) \not\approx \mathcal{O}_{\mathbf{P}^n}(p)$. $\square$

We will later discuss smooth varieties. For them we have

Theorem 5.5.5. *If X is a smooth variety then all the local rings $\mathcal{O}_{X,x}$ are UFDs.*

This theorem is best proved by syzygies à-la Auslander–Buchsbaum [AB]; a geometric argument is given in Mumford [M2-1]. Several algebraic proofs are in Zariski-Samuel [ZS] but we won't be using this theorem in any essential way.

5.6 Operations on sheaves that change spaces

Let $f : X \to Y$ be a continuous mapping of topological spaces. Let $\mathcal{F}$ and $\mathcal{G}$ be sheaves on X and Y. Then an *f-homomorphism* $\psi : \mathcal{G} \to \mathcal{F}$ is a collection of operators $\psi_U : \mathcal{G}(U) \to \mathcal{F}(f^{-1}U)$ for each open subset $U \subset \psi$ taking a section α of $\mathcal{G}$ to the section $\psi(\alpha) = \psi_U(\alpha)$ such that $\psi(\alpha|_V) = \psi(\alpha)|_{f^{-1}V}$ for all open subsets V of U. In the picture we have

$$\left\{ \begin{array}{c} \mathcal{F} \xleftarrow{\psi} \mathcal{G} \\ X \to Y \\ f \end{array} \right\}$$

If $\mathcal{F}$ and $\mathcal{G}$ are abelian (sheaves of rings or k-algebras) we define ψ to be a homomorphism if each ψ_U is.

We have many examples of this concept. If $f = 1_X$, ψ is just an ordinary mapping of sheaves. If $f : X \to Y$ is a morphism of spaces with functions then f^* defines a k-algebra f-homomorphism $\mathcal{O}_Y \to \mathcal{O}_X$. If X is a closed or open subvariety of a variety Y then for any sheaf $\mathcal{F}$ of $\mathcal{O}_Y$-modules, we have an obvious i-homomorphism $\mathcal{F} \to \mathcal{F}|_X = i^*\mathcal{F}$ where $i : X \to Y$ is the inclusion.

Let $f : X \to Y$ be a continuous mapping. Let $\mathcal{F}$ be a sheaf on X. We can define a sheaf $f_*\mathcal{F}$ on Y by the rule $V \mapsto \mathcal{F}(f^{-1}V)$ with the obvious sections. Then we have a tautological f-homomorphism $\rho : f_*\mathcal{F} \to \mathcal{F}$. Composition with ρ defines a bijection

$$\mathrm{Hom}_Y(\mathcal{G}, f_*\mathcal{F}) = f\text{-}\mathrm{Hom}(\mathcal{G}, \mathcal{F})$$

for sheaves $\mathcal{G}$ on Y. Clearly $f_*\mathcal{F}$ is abelian if $\mathcal{F}$ is. We have an adjoint which assigns a sheaf $f^{-1}\mathcal{G}$ on X to a sheaf $\mathcal{G}$ on Y with a f-homomorphism $\sigma : \mathcal{G} \to f^{-1}\mathcal{G}$. In this case composition with σ defines a bijection

$$\mathrm{Hom}_X(f^{-1}\mathcal{G}, \mathcal{F}) = f\text{-}\mathrm{Hom}(\mathcal{G}, \mathcal{F})$$

for all sheaves $\mathcal{F}$ on X. As for a construction $f^{-1}\mathcal{G}$ is the sheaf associated to the presheaf

$$U \to \mathrm{limit}\ \mathcal{G}(V).$$
$$V \subset Y \text{open}$$
$$V \supseteq f(U)$$

Fortunately this hard concept is not used too much in algebraic geometry.

Assume furthermore we have an f-homomorphism $f^* : \mathcal{A}_Y \to \mathcal{A}_X$ where the $\mathcal{A}_X$ and $\mathcal{A}_Y$ are sheaves of rings on X and Y. Then if $\psi : \mathcal{G} \to \mathcal{F}$ is an f-homomorphism where $\mathcal{F}$ and $\mathcal{G}$ are $\mathcal{O}_X$ and $\mathcal{O}_Y$-modules then ψ is an (f, f^*)-*homomorphism* if $\psi(g \cdot \alpha) = f^*g \cdot \psi(\alpha)$ where g is a local section of $\mathcal{O}_Y$ and α is a local section of $\mathcal{G}$. Clearly if $\mathcal{F}$ is any $\mathcal{O}_X$-module then $f_*\mathcal{F}$ has a unique structure of an $\mathcal{A}_Y$-module such that $\rho : f_*\mathcal{F} \to \mathcal{F}$ is an (f, f^*)-homomorphism. We have a bijection $\mathcal{A}_Y\text{-}\mathrm{Hom}(\mathcal{G}, f_*\mathcal{F}) = (f, f^*)\text{-}\mathrm{Hom}(\mathcal{G}, \mathcal{F})$ for any $\mathcal{O}_Y$-module $\mathcal{G}$.

In the other direction given an $\mathcal{O}_Y$-module $\mathcal{G}$ we may have to define an $\mathcal{A}_X$-module $f^*\mathcal{G}$ with an (f, f^*)-homomorphism $\sigma' : \mathcal{G} \to f^*\mathcal{G}$ such that $\mathcal{A}_X\text{-}\mathrm{Hom}(f^*\mathcal{G}, \mathcal{F}) = (f, f^*)\text{-}\mathrm{Hom}(\mathcal{G}, \mathcal{F})$ for all $\mathcal{A}_X$-modules $\mathcal{F}$. $f^*\mathcal{G}$ is the sheaf associated to the presheaf $U \mapsto f^{-1}\mathcal{G}(U) \otimes_{f^{-1}} \mathcal{O}_Y(U)\mathcal{O}_X(U)$. In algebraic geometry f^* is trivial to compute and f_* is difficult contrary to their definitions.

Let $f : X \to Y$ be a morphism of varieties. Then we have the implicit f-homomorphism $f^* : \mathcal{O}_Y \to \mathcal{O}_X$ of k-algebras.

Lemma 5.6.1.

(a) If $\mathcal{G}$ is a (quasi-)coherent $\mathcal{O}_Y$-module then $f^*\mathcal{G}$ is a (quasi-)co-herent $\mathcal{O}_X$-module. Also $f^*\mathcal{G}$ is locally free or invertible if $\mathcal{G}$ is.

(b) If $\mathcal{F}$ is a quasi-coherent $\mathcal{O}_X$-module then $f_*\mathcal{F}$ is a quasi-coherent $\mathcal{O}_Y$-module.

(c) f^* is right exact and f_* is left exact and both are additive and commute with direct limits.

Proof. We begin with (c). f^{-1} is clearly exact as the stalk of $f^{-1}\mathcal{G}$ at a point x = stalk of $\mathcal{G}$ at $f(x)$. For f^* is right exact because tensoring is right exact. Clearly f^* commutes with direct limits because tensoring does. For f_* let $0 \to \mathcal{F}_1 \to \mathcal{F}_2 \to \mathcal{F}_3 \to 0$. Then $0 \to f_*\mathcal{F}_1 \to f_*\mathcal{F}_2 \to f_*\mathcal{F}_3$ is even exact as a sequence of presheaves much less sheaves. For direct limit, we need $(\underrightarrow{\lim} f_*\mathcal{F}_i)(V) = \underrightarrow{\lim}(\mathcal{F}_i(f^{-1}V))$ but this is true because spaces are noetherian. Thus (c) is true.

For (a) as the statement is locally on X and Y we assume that X and Y are affine. Let M be a $k[Y]$-module. We have the homomorphism $f^* : k[Y] \to k[X]$.

Claim. $f^*(\tilde{M}) = (M \otimes_{k[Y]} k[X])^\sim$.

This claim clearly implies (a).

For the claim let $\psi : k[Y]^{\oplus I} \to k[Y]^{\oplus J} \to M \to 0$ be the presentation of M. Then $\psi : \mathcal{O}_Y^{\oplus I} \to \mathcal{O}_Y^{\oplus J} \to \tilde{M} \to 0$ is exact. By (c) $f^*\psi : \mathcal{O}_X^{\oplus I} \to \mathcal{O}_X^{\oplus J} \to f^*(\tilde{M}) \to 0$ is exact. Thus $f^*(\tilde{M}) = (\text{Cok } \Gamma(X, f^*\psi))^\sim$ but $\Gamma(X, f^*\psi)$ is the matrix $(f^*\psi_{i,j})$ where $\psi = (\psi_{i,j})$. Thus Cok $\Gamma(X, f^*\psi) = M \otimes_{k[Y]} k[X]$.

For (b) as the statement is local on Y, let g be a regular function on Y. We need to prove that

$$\underrightarrow{\lim}(\frac{1}{g^i} f_*\mathcal{F}) \overset{\approx}{\to} _{D(g)} f_*\mathcal{F}$$

is an isomorphism. As $\mathcal{F}$ is quasi-coherent

$$\underrightarrow{\lim} \left(\frac{1}{(f^*g)^i} \mathcal{F}\right) \overset{\approx}{\to} _{D(f^*g)}\mathcal{F}$$

is an isomorphism by Proposition 5.2.2. So by (c),

$$\underrightarrow{\lim}(\frac{1}{g^i} f_*\mathcal{F}) \overset{\approx}{\to} f_*(_{D(f^*g)}\mathcal{F}) = _{D(g)} f_*\mathcal{F}$$

where the last equation is trivial. Thus (b) is true. $\square$

5.7   Morphisms to projective space and affine morphisms

Let $f : X \to \mathbf{P}^n$ be a morphism where X is a variety. Then we have the basic sections $X_0, \ldots, X_n$ of $\mathcal{O}_{\mathbf{P}^n}(1)$. Pulling-back we get sections $\sigma_0, \ldots, \sigma_n$ of the invertible sheaf $\mathcal{L} = f^* \mathcal{O}_{\mathbf{P}^n}(1)$. Then the image of the germs of $\sigma_0, \ldots, \sigma_n$ span the $\mathcal{O}_{X,x}$-module $\mathcal{L}|_x$ at each point x of X since $\sigma_1, \ldots, \sigma_n$ span $\mathcal{L}$.

Lemma 5.7.1. *Conversely, suppose we are given an invertible sheaf $\mathcal{L}$ on X and sections $\sigma_0, \ldots, \sigma_n$ which span $\mathcal{L}$. Then there is a unique morphism $f : X \to \mathbf{P}^n$ such that $\mathcal{L}$ is naturally isomorphic to $f^* \mathcal{O}_{\mathbf{P}^n}(1)$ and the σ_0's correspond to the previous σ_i.*

Proof. Let $D_i = \{x \in X | \sigma_i(x) \neq 0\}$. Then the D_i give an open cover of X as the σ_i's span $\mathcal{L}$. We define $f_i(x) = (\sigma_0/\sigma_i)(x), \ldots, (\sigma_n/\sigma_i)(x)$ for x in $D_i s$) where the ratios σ_j/σ_i are regular functions on D_i. Then one checks that f_i and f_j agree on $D_i D_j$. Thus they give a morphism $f : X \to \mathbf{P}^n$. Then one checks that this gives the only solution to the problem. $\qquad\square$

Exercise 5.7.2.   Give the details of the last proof.

An invertible sheaf $\mathcal{L}$ on a variety X is called *very ample* if it has sections $\sigma_0, \ldots, \sigma_n$ which define a morphism $f : X \to \mathbf{P}^n$ as before such that X is isomorphic to its image which is a subvariety of $\mathbf{P}^n$. Such $\mathcal{L}$ is just *ample* if $\mathcal{L}^{\otimes n}$ is very ample for some positive n.

Exercise 5.7.3.   Show that $\mathcal{O}_X$ is very ample if X is an affine variety.

Exercise 5.7.4.   Let X be a product $X_1 \times X_2$. Let $\mathcal{L}_1$ and $\mathcal{L}_2$ be ample invertible sheaves on X_1 and X_2. Then $\pi_{X_1}^* \mathcal{L}_1 \otimes \pi_{X_2}^* \mathcal{L}_2$ is ample on X.

Exercise 5.7.5.   Let $\mathcal{L}$ and $\mathcal{M}$ be invertible sheaves on a variety X. Assume that $\mathcal{L}$ is generated by its sections; i.e. there are sections $\sigma_1, \ldots, \sigma_n$ of $\mathcal{L}$ such that zeroes$(\sigma_1, \ldots, \sigma_n) = 0$. Then if $\mathcal{M}$ is ample then $\mathcal{L} \otimes \mathcal{M}^{\otimes n}$ is ample for some $n \geq 0$.

Exercise 5.7.6.   Let $\mathcal{L}$ be an invertible sheaf on a projective variety $X \subset \mathbf{P}^n$. Then $\mathcal{L}(m)$ is very ample for $m \gg 0$.

We are now in a position to understand affine and finite morphisms.

Lemma 5.7.7.

(a)   *A morphism $f : X \to Y$ of varieties is affine iff for any open affine subset V $f^{-1}V$ is affine.*

(b)   *An affine morphism is finite iff $k[f^{-1}V]$ is a $k[V]$-module of finite type for all such V.*

Proof. By definition f is affine or finite if there is an open affine cover $Y = \bigcup V_i$ such that V_i has the required property. Thus the "if" is trivial. The other way is a local to global argument. As the problem is local on Y we may assume that Y is affine and equals V. Thus for (a) we want to show that X is affine. Let $A = k[X] = \Gamma(Y, f_*\mathcal{O}_X)$. Let $Y = \bigcup D(g_i)$ be a finite covering of Y where the g_i's are contained in $k[Y]$ such that $f^{-1}D(g_i)$ is affine for each i. Then as $f_*\mathcal{O}_X$ is quasi-coherent then $A_{(g_i)} = \Gamma(D(g_i), f_*\mathcal{O}_X) = k[f^{-1}D(g_i)]$. Now A is an algebra of functions. So it has no nilpotent. Also A is a finitely generated k-algebra because its finite number of localizations $A_{(g_i)}$ are (why?). So we have morphism $h : X \to \operatorname{Spec} A \xrightarrow{\pi} \operatorname{Spec} k[Y] = Y$ with composition f. We know that the $h : f^{-1}D(g_i) \to \pi^{-1}D(g_i)$ are all isomorphisms. So h is an isomorphism. Thus (a) is true. For (b) we have $A_{(g_i)}$ is a finite type $k[Y]_{(g_i)}$-module $\Rightarrow f_*\mathcal{O}_X$ is coherent $\Rightarrow A$ is a finitely generated $k[Y]$-module. $\qquad\qquad\square$

6

Smooth varieties and morphisms

6.1 The Zariski cotangent space and smoothness

Let x be a point in a variety X. We have the local ring $\mathcal{O}_{X,x}$ of X at x and its maximal ideal m_x. The quotient field $\mathcal{O}_{X,x}/m_x$ is k and the corresponding surjection $\mathcal{O}_{X,x} \to k$ is just given by evaluating a germ of a function at x. We want to study the first order variation of a function at x. Consider the k-vector-space m_x/m_x^2. This finite dimensional space $\mathrm{Cot}_x(X)$ is called the *Zariski cotangent space* of X at x.

Let f be a germ in $\mathcal{O}_{X,x}$. We define the *differential* $df|_x$ of f at x to be the class of $f - f(x)$ in m_x/m_x^2. Thus we have a mapping $d{-}|_x : \mathcal{O}_{X,x} \to \mathrm{Cot}_x(X)$ which satisfies the following properties:

(a) d constant $|_x = 0$
(b) $d(f + g)|_x = df|_x + dg|_x$ and
(c) $d(f \cdot g)|_x = f(x)dg|_x + g(x)df|_x$.

Exercise 6.1.1. Prove these properties. Furthermore show that any mapping $\psi : \mathcal{O}_{X,x} \to W$ into a k-vector-space with the same properties has the form $\lambda(df|_x) = \psi(f)$ where $\lambda : m_x/m_x^2 \to W$ is a uniquely determined k-linear mapping.

Clearly the differential $df|_x$ describes the first order variation of f at x. The first properties of Cot_x are given in

Lemma 6.1.2.

(a) $\dim_k \mathrm{Cot}_x X \geq \dim_x X \equiv \max\{\dim C\}$, *where C is a component of X passing through x.*

(b) $\mathrm{Cot}_{(x_1, x_2)}(X_1 \times X_2) = \mathrm{Cot}_{x_1}(X_1) \oplus \mathrm{Cot}_{x_2}(X_2)$.

(c) *If X is an affine variety and n_x is the maximal ideal of regular functions vanishing at x then $\mathrm{Cot}_x X = n_x / n_x^2$.*

Proof. To prove (a) we may assume by replacing X by a smaller open set that X is affine and $\dim_x X = \dim X$. Let $f_1, \ldots, f_m$ be regular functions on X which vanish at x such that the differentials $df_1|_x, \ldots, df_m|_x$ span $\mathrm{Cot}_x X$. We want to show that $m \geq \dim X$.

Let $Y = \mathrm{zeroes}(f_1, \ldots, f_m)$ in X. Then x is in Y. Let $\mathcal{I}$ be the ideal sheaf of $\{x\}$ in X. Then $(f_1, \ldots, f_m) \subset \mathcal{I}$. By assumption $(f_1, \ldots, f_m)$ $|_x \to \mathcal{I}|_x$ is surjective. Thus $(f_1, \ldots, f_m) = \mathcal{I}$ in a neighborhood of x. Thus x is an isolated point (i.e. component) of Y. By the corollary to the principal ideal theorem $\dim(X) - m \leq \dim\{x\} = 0$. This proves (a).

For (c), as any element in $k[X] - n_x$ acts invertibly on n_x/n_x^2, (n_x/n_x^2) $= (n_x/n_x^2)_{n_x} = (n_{x_{n_x}}/n_{x_{n_x}}^2) = m_x/m_x^2$.

For (b) we may assume that X_1 and X_2 are affine. By (c),

$$\mathrm{Cot}_{(x_1, x_2)}(X_1 \times X_2) = n_{x_1} \otimes k[X_2] + k[X_1] \otimes n_{x_2}/_{(\text{same})^2}$$

$$= n_{x_1}/n_{x_1}^2 \otimes k \oplus k \otimes n_{x_2}/n_{x_2}^2$$

$$= \mathrm{Cot}_{x_1}(X_1) \oplus \mathrm{Cot}_{x_2}(X_2).$$

$\square$

The Zariski cotangent space is an algebraic object. We may want to define a geometric object $T_x X$ which is the *tangent space* of X. By definition $T_x X$ is the affine space whose space of linear functionals is $\mathrm{Cot}_x X$. Thus $T_x X = \mathrm{Spec}(\bigoplus_{n \geq 0} \mathrm{Sym}^n(\mathrm{Cot}_x X))$. Clearly $T_x X$ has the same dimension as $\mathrm{Cot}_x X$.

A variety X is *smooth* at point x if $\dim T_x X = \dim_x X$. The variety is smooth if it is smooth at all its points. In Section 6.2 we will prove that if X is smooth at x then X is irreducible in a neighborhood of x. In Section 6.3 we will prove that the set of points where a variety is smooth is open. In this section we will see that this subset is non-empty. To do this we may assume that the variety is irreducible by looking for such points not on any other component.

Lemma 6.1.3. *If X is a non-empty irreducible variety, then X is smooth on an open dense subset.*

Proof. We may assume that X is affine or, even, X is closed in $\mathbf{A}^n$ where $n = \dim X + d$. If $d = 0$ there is no problem because $X = \mathbf{A}^n$ where $\mathbf{A}^n$ is clearly smooth at all its points. We use induction on d. If $d > 0$ we may find a non-constant regular function g on $\mathbf{A}^n$ such that g has minimal degree such that $X \subset (g = 0)$.

Claim. Some partial derivative $\frac{\partial g}{\partial X_i}$ does not vanish identically on X.

Proof of Claim. Otherwise all $\frac{\partial g}{\partial X_i}$ are identically zero because their degree $< \deg g$. In characteristic zero this means that g is constant $\neq 0$ but X is non-empty. This is impossible. In characteristic p, we have $g = \sum a_{(r)} X^{p(r)}$. Thus $g^{1/p} = \sum a_{(r)}^{1/p} X^{(r)}$ is a regular function vanishing on X of smaller degree than g (unless g is constant). This is also impossible.

We may assume that $\frac{\partial g}{\partial X_n} \neq 0$ and now we use the preparation lemma, which we know works for $\mu_1, \ldots, \mu_n$ in an open dense subset of $\mathbf{A}^n$. Thus if we make a general change of coordinates g will be monic in X_n and $\frac{\partial g}{\partial X_n} \neq 0$ on X. Consider the projection

$$X \subset (g = 0) \subset \mathbf{A}^n.$$
$$\searrow \quad \downarrow \quad \swarrow$$
$$\mathbf{A}^{n-1}$$

As usual π is finite. Let Y be the image of $\pi(X)$. Then as $\dim X = \dim Y$ we know the lemma for Y. Let x be a point of X such that $\frac{\partial g}{\partial X_n}|_x \neq 0$ and $\pi(x)$ is a point of Y at which Y is smooth. We have a surjection $\mathrm{Cot}_{\pi(x)}(Y) \oplus k dX_n|_x /(dg|_x) \to \mathrm{Cot}_x(X)$ where $dg|_x \neq 0$. Thus $\dim \mathrm{Cot}_x X \leq \dim \mathrm{Cot}_{\pi(x)} Y = \dim Y = \dim X$. Thus $\dim \mathrm{Cot}_x X = \dim X$ by Lemma 6.1.1(a). Hence X is smooth at x. Clearly the $\{x\}$ contain an open dense subset. $\square$

6.2 Tangent cones

In this section we will study higher order terms. Let x be a point on a variety X. By definition $G_x(\mathcal{O}_{X,x})$ is the graded ring $\bigoplus_{n \geq 0} (m_x^n / m_x^{n+1})$. Then $G_x(\mathcal{O}_{X,x})$ contains $\mathrm{Cot}_x(X)$ as its degree one term. Furthermore $G_x(\mathcal{O}_{X,x})$ is generated by $\mathrm{Cot}_x(X)$ as a k-algebra. Thus we have a surjection $\psi : \bigoplus_{n \geq 0} \mathrm{Sym}^n(\mathrm{Cot}_x(X)) \to G_x(\mathcal{O}_{X,x})$.

To make some geometry out of this we have the cone $(TC_x X)_{\mathrm{red}} = $ zeroes $(\mathrm{Ker}\ \psi) \subset T_x X$. The TC stands for *tangent cone* and red means we are giving it the wrong scheme structure because we don't know schemes. Thus

$$(\mathrm{TC}_x X)_{\mathrm{red}} = \mathrm{Spec}(G_x(\mathcal{O}_{X,x})/\sqrt{0}).$$

In this section we will give a geometric interpretation of the tangent cone. For this study which is local we will assume that X is affine and we have a closed embedding $X \subset \mathbf{A}^n$ such that x is the origin 0.

Proposition 6.2.1. $(\mathrm{TC}_x X)_{\mathrm{red}} = $ *union of all limiting secants to X at x.*

First we need to give a precise meaning to this statement. Let y be a point of $X - \{0\}$. The *secant* ℓ_y is the line spanned by x and y. The limit ℓ_y as $y \to x$ in $\mathbf{P}^{n-1}$ should be a limiting secant. We will make the limiting process precise as follows. The pair (y, ℓ_y) are points of $\mathbf{A}^n$ blown up at the origin $B_0(\mathbf{A}^n)$. They are exactly $\pi^{-1}(X - \{0\})$ where $\pi : B_0(\mathbf{A}^n) \to \mathbf{A}^n$ is the projection that sends $(z \in m)$ to z. Then the limiting secants are $K = \overline{\pi^{-1}(X - \{0\})} \cap (\text{exception divisor } E = \{0 \in \ell\} = (0 \times \mathbf{P}^{n-1}))$. The precise statement of the proposition is simply

$(*)$ $\qquad\qquad (\mathrm{TC}_x X)_{\mathrm{red}}$ is the cone $C(K)$ over K.

We need to show that $\pi'((\mathrm{TC}_x X)_{\mathrm{red}} - (0)) = K$ where $\pi' : \mathbf{A}^n - \{0\} \to \mathbf{P}^{n-1}$ is a projection. This question is local on $\mathbf{P}^{n-1}$. So we may compute in $D_i = \{(x) \in \mathbf{P}^{n-1} | x_i \neq 0\} = \{(y_1, \ldots, 1(i\text{-place}), \ldots, y_n)\}$. A point in $B_0(\mathbf{A}^n) \cap \mathbf{A}^n \times D_i = R_i$ has the form $(\lambda y, y)$ for some λ in D_i. Thus (λ, y) are coordinates there. Now $E \cap R_i$ is given by the one equation $\lambda = 0$. We have $\pi^{-1}(X - \{0\}) \cap R_i = \{(\lambda y, y) | \lambda \neq 0$ and $f(\lambda \cdot y) = 0$ for all f in $k[\mathbf{A}^n]$ vanishing on $X\}$. So $\pi^{-1}(X - \{0\}) \cap R_i = \{(\lambda y, y) | g(\lambda, y) = 0$ where $\lambda^N g(\lambda, y) = f(\lambda \cdot y)$ for some $N > 0\} = \{g(\lambda, y) = 0 | g(\lambda, y) = \frac{f(\lambda \cdot y)}{\lambda^N}$ where $N = \max i$ such that $f \in (X_1, \ldots, X_n)^N \cap I$ where I is the ideal of X in $\mathbf{A}^n\}$. So $K \cap R_i = \{(0, y) | f_N(y) = 0\}$ where $f = f_N + $ higher order terms with f in I and $f_N \neq 0$ so zeroes$\{\mathrm{Ker}(\psi)\} = \pi'(\mathrm{TC}_x X)_{\mathrm{red}} - (0) \cap R_i$. This proves the proposition. $\qquad\square$

As a consequence we have

Corollary 6.2.2. *If x is not an isolated point of X, the dimension of any component of $(\mathrm{TC}_x X)_{\mathrm{red}}$ equals the dimension of some component of X passing through x.*

Proof. Clearly we may assume that each component of X passes through x. The components of $\pi^{-1}(X - \{0\})$ have the form $\pi^{-1}(C - \{0\})$ where

C is a component of X. As E is locally defined by one equation, $\overline{\pi^{-1}C - \{0\}} \cap E$ has components all of whose dimension is dim $C - 1$. By the proposition these correspond to components of $TC_x(C)$ which has dimension dim C. Again we have $TC_x(X)_{\mathrm{red}} = \bigcup TC_x(C)_{\mathrm{red}}$. Thus the corollary is true. $\qquad\qquad\qquad\qquad\qquad\qquad\qquad\qquad\qquad\qquad\square$

We may now prove

Lemma 6.2.3. *If x is a point of a variety X at which X is smooth, then X has only one of its components passing through x.*

Proof. We may assume that $\dim_x X = \dim\ X$. Now $\dim\ TC_x X_{\mathrm{red}} \le \dim\ T_x X = \dim\ X$ but these are the same by the corollary. Hence $TC_x X_{\mathrm{red}} = T_x X$. Thus $G_x(\mathcal{O}_{X,x})$ is the polynomial ring $\mathrm{Sym}(\mathrm{Cot}_x X)$. Hence $G_x(\mathcal{O}_{X,x})$ is an integral domain. Therefore if $f \in m_x^i - m_x^{i+1}$ and $g \in m_x^j - m_x^{j+1}$ then $fg \in m_x^{i+j} - m_x^{i+j+1}$. Thus $\mathcal{O}_{X,x}$ will be an integral domain if we can prove

Sublemma 6.2.4. $\bigcap\limits^{i} m_x^i = 0$.

Proof. Assume X is affine $\subseteq \mathbf{A}^n$ where $x = 0$. Let $\tilde{\pi} : \tilde{X} \to X$ be the blowing up of X where $\tilde{X} = \overline{\pi^{-1}(X - \{0\})}$ as before. Then $\tilde{\pi}^{-1}(x) = K$ is locally defined by one equation in $\tilde{X}$. Let k be a point of K and $e = 0$ be a local equation for E near k. Pulling back $(\tilde{\pi}^*)(\bigcap\limits^{i} m_x^i) \subseteq \bigcap\limits^{i}(e^i \mathcal{O}_{\tilde{X},k})$. If we show that this last ideal is zero for all i then $\tilde{\pi}^* f = 0$ in a neighborhood U of K if $f \in \bigcap\limits^{i} m_x^i$. Hence, if X is not a point, $\tilde{\pi}$ is surjective and $\tilde{\pi}(U)$ contains a neighborhood of x because $\mathbf{P}^{n-1}$ is complete. We can now finish if we prove the general

Claim. If e is an element of the maximal ideal of a noetherian local ring A, then $\bigcap\limits^{i} e^i A = 0$.

Proof. Let a be an element of the intersection. Let $[Aa : e^i A] = \{b \in A | e^i b \in Aa\}$. Thus we have an increasing sequence $[Aa : e^i A] \subseteq [Aa : e^{i+1} A]$. As A is noetherian $[Aa : e^n A] = [Aa : e^{n+1} A] = I$ for some n. By assumption $Aa = [Aa : e^i A] \cdot e^i A$. Thus $I \cdot e^n = I \cdot e^{n+1}$. Therefore by Nakayama's lemma $I \cdot e^n = 0$ but $a \in I \cdot e^n$. So $a = 0$. $\qquad\square$

Exercise 6.2.5. Compute the tangent cone and blowing up of $\mathbf{A}^2$ at 0 in

(a) $x = y^2$ (parabola),
(b) $x^2 = y^3$ (cusp),
(c) $(x + y)(x - y) + y^3 = 0$ (node).

6.3 The sheaf of differentials

Let X be a variety. We have defined the coherent sheaf Ω_X on X as the sheaf corresponding to the sheaf $\mathcal{I}_\Delta / \mathcal{I}_\Delta^2$ where $\Delta \subset X \times X$ is the diagonal. In this section we will learn how to compute Ω_X. The first fact is that we have a k-homomorphism $d : \mathcal{O}_X \to \Omega_X$ which sends a local regular function g to the differential corresponding to the $(g(x_1) - g(x_2))$-modulo $\mathcal{I}_\Delta^2$.

We will compute when X is affine. Then $\Omega_X = \widetilde{\Omega[X]}$ where $\Omega[X]$ is the $k[X]$-module I/I^2 where I is the kernel of the multiplication $k[X] \otimes k[X] \to k[X]$. Then we have $d : k[X] \to \Omega[X]$.

Lemma 6.3.1.

(a) $d(\text{constant}) = 0$.
(b) $d(f + g) = df + dg$ and $d(f \cdot g) = f \cdot dg + g \cdot df$ for any f and g in $k[X]$.
(c) $\Omega[X]$ is generated by the differentials df as a $k[X]$-module.
(d) Given any $k[X]$-module M and a mapping $\delta : k[X] \to M$ satisfying the same formal rules as d, there is a unique $k[X]$-module homomorphism $\ell : \Omega[X] \to M$ such that $\delta(f) = \ell(df)$ for all regular functions f on X.

Proof. (a) and (b) are obvious. For instance, the last part of (b) follows from the identity

$$f(x_1)g(x_1) - f(x_2)g(x_2) = f(x_1)(g(x_1) - g(x_2)) + g(x_2)(f(x_1) - f(x_2)).$$

Part (c) is a consequence of the fact that $f(x_1) - f(x_2)$ generate the ideal of Δ_X. This is a special case of Lemma 3.3.2(d) as Δ_X is the graph of the identity of X.

The uniqueness in (d) follows from (c). For the existence, first note that the rules imply that $\delta : k[X] \to M$ is a k-linear mapping. Thus we may extend δ to a $k[X]$-module homomorphism $\nabla : k[X \times X] \approx k[X] \otimes_k k[X] \to M$, where $k[X \times X]$ is regarded as a $k[X]$-module via

π_2^*. Explicitly, $\nabla(\sum f_i(x_1)g_i(x_2)) = \sum g_i(x) \cdot \delta f_i$. One easily checks the rules:

(1) $\nabla(f(x_1)) = \delta f$,

(2) $\nabla(f(x_2)) = 0$,

(3) $\nabla(f(x,y) + g(x,y)) = \nabla(f(x,y)) + \nabla(g(x,y))$ and

(4) $\nabla(f(x,y)g(x,y)) = f(x,x) \cdot \nabla(g(x,y)) + g(x,x) \cdot \nabla(f(x,y))$.

Using rule (4), you will find $\nabla(I^2) = 0$. By the other rules, $\nabla(f(x_1) - f(x_2)) = \delta f$. Thus ∇ induces a mapping $\ell : I/I^2 = \Omega[X] \to M$ such that $\ell(df) = \delta f$. By rules (2) and (4), ℓ is a $k[X]$-module homomorphism. Thus the required ℓ exists.   □

Example. Let $X_1, \ldots, X_n$ be the coordinate functions on $\mathbf{A}^n$. Then $\Omega[\mathbf{A}^N]$ is the free $k[\mathbf{A}^N]$-module with basis the differentials $dX_1, \ldots,$ dX_n. The classical proof of this will be given in vector notation. We have an isomorphism $i : \mathbf{A}^n \times \mathbf{A}^n \to \mathbf{A}^n \times \mathbf{A}^n$ given by $i(\mathbf{x}, \delta) = (\mathbf{x}, \mathbf{x} + \delta)$. Clearly, i induces an isomorphism $\mathbf{A}^n \times \{0\}$ with $\Delta_{\mathbf{A}^n}$. Let J be the ideal of $\mathbf{A}^n \times \{0\}$ in $\mathbf{A}^n \times \mathbf{A}^n$. Evidently, J/J^2 is a free $k[X_1, \ldots, X_n]$-module with basis the classes of $\delta_1, \ldots, \delta_n$. Using the isomorphism i, we see that we have computed $\Omega[X]$.

Let $f : X \to Y$ be a morphism between affine varieties. Then $(f \circ \pi_1, f \circ \pi_2) : X \times X \to Y \times Y$ is a morphism, which takes Δ_X into Δ_Y. Its comorphism must induce an f^*-homomorphism, $f^* : \Omega[Y] = I_Y/I_Y^2 \to \Omega[X] = I_X/I_X^2$. This means that

(1) $f^*(a \cdot \omega) = f^*(a) \cdot f^*(\omega)$ for $a \in k[Y]$ and $\omega \in \Omega[Y]$.

Furthermore, from the definition, one easily verifies the equation

(2) $d_X(f^*a) = f^*(d_Y a)$ for $a \in k[Y]$.

In general, f^* induces a $k[X]$-homomorphism $f' : \Omega[Y] \otimes_{k[Y]} k[\ell] \to \Omega[X]$. The next lemma will explain (in a particular case) how f' may be used to compute $\Omega[X]$ for an arbitrary affine X as X is isomorphic to a closed subvariety of an affine space. The general principle is called *implicit differentiation* in calculus books.

Lemma 6.3.2. *Let Z be a closed subvariety of an affine variety X. Let $J = \mathrm{ideal}(Z) \subset k[X]$. Then we have an exact sequence of $k[Z]$-modules,*

$$J/J^2 \xrightarrow{\mathrm{d}} \Omega[X] \otimes_{k[X]} k[Z] \xrightarrow{i'} \Omega[Z] \to 0,$$

where i is the inclusion $Z \subset X$ and $\mathrm{d}(a) = d_X(a) \otimes 1$ for any a in J.

Proof. By Lemma 6.3.1(c), i' is surjective as $i^* : k[X] \to k[Z]$ is

surjective. As $i^*(d_X a) = d_Z(i^* a) = d_Z(0) = 0$ when $a \in J, i^* d_X(J) = 0$.
Thus $i' \circ d : J \to \Omega[Z]$ is zero. By the product formula for differentiation,
$d_X(J^2) \subseteq J \cdot \Omega[X]$. Hence d induces the $k[Z] = k[X]/J$-homomorphism,
$\mathbf{d} : J/J^2 \to \Omega[X] \otimes_{k[X]} k[Z]$, mentioned in the statement. The remaining
question is the exactness in the middle.

Is the induced $k[Z]$-homomorphism $m : M = \Omega[X] \otimes_{k[X]} k[Z]/\mathbf{d}(J/J^2)$
$\to \Omega[Z]$ an isomorphism? As m is surjective, we may answer "yes" if we
find a surjective $k[Z]$-homomorphism $\ell : \Omega[Z] \to M$ such that $\ell \circ m$ is
the identity of M. Let a be a regular function on Z. Let a' be a lifting
of a to $k[X]$. Note that $d_X(a')$ is determined by a modulo$(d_X J)$. Let
$\delta(a)$ be the class of $d_X(a')$ in M. Obviously, $\delta : k[Z] \to M$ satisfies the
same formal rules for differentiation as d_X does. By Lemma 6.3.1(d), we
have a $k[Z]$-homomorphism $\ell : \Omega[X] \to M$ such that $\ell(d_Z(a)) = \delta(a)$.
By (2) above, $\ell \circ m(d_X(a)) = d_Z(a)$. The other requirement for ℓ that
we want follows because the $\delta(a)$ generate M and the $d_Z(a)$ generate
$\Omega[Z]$ by Lemma 6.3.1. □

We have thus far defined differentials and noted some of their formal
properties, but we have failed in relating df with the change of f at a
point x of our affine variety X. This void will be filled in the remainder
of the section.

Lemma 6.3.3. *We have a natural isomorphism*
$$\Omega_X|_x = \mathrm{Cot}_x(X).$$

Proof. We may assume that X is affine and n is a maximal ideal of x
in $k[X]$. Then we want a natural isomorphism
$$\Omega[X] \otimes_{k[X]} k[X]/n \approx n/n^2.$$
We have the mapping $d-|_x : k[X] \to n/n^2$ which satisfies the usual rules.
Thus by Lemma 6.3.1 there is a unique $k[X]$-linear mapping $\ell : \Omega[X] \to$
n/n^2 such that $df|_x = \ell(df)$. As multiplication by n annihilates n/n^2, ℓ
induces a k-linear mapping
$$\bar{\ell} : \Omega[X] \otimes_{k[X]} k[X]/n \to n/n^2 \text{ such that } df|_x = \bar{\ell}(df \otimes 1).$$
Now the mapping $k[X] \to \Omega[X] \otimes_{k[X]} k[X]/n$ sending f to $df \otimes 1$ satisfies
the same formal rules as $d-|_x$. Thus there is a unique linear mapping
$m : n/n^2 \to \Omega[X] \otimes_{k[X]} k[X]/n$ such that $m(df|_x) = df \otimes 1$. Clearly $\bar{\ell}$
and m are inverse of each other. □

Exercise 6.3.4. Find the Zariski cotangent space of $C = \{x^2 = y^3\}$ at
each of its points.

Exercise 6.3.5. Let $f(X_1, \ldots, X_n)$ be a regular function on $\mathbf{A}^n$. Show that $df = \frac{\partial f}{\partial X_1} dX_1 + \cdots + \frac{\partial f}{\partial X_n} dX_n$ where the partial derivatives are computed as usual (for polynomials).

Exercise 6.3.6. Let $X \times Y$ be the product of two affine varieties. Show that there is a natural isomorphism between $\Omega[X \times Y]$ and $\Omega[X] \otimes_{k[X]} k[X \times Y] \oplus \Omega[Y] \otimes_{k[Y]} k[X \times Y]$. In other words, any regular differential on $X \times Y$ can be written $f(y) \cdot \omega_X + g(x) \cdot \omega_Y$ for $\omega_X \in \Omega[X]$ and $\omega_Y \in \Omega[Y]$, where ω_X and ω_Y are unique up to constant factors.

Exercise 6.3.7. Let C be the circle $x^2 + y^2 = 1$. Show that

(a) any regular function on C may be written uniquely $f[Y] + Xg[Y]$, where f and g are polynomials,

(b) any regular differential on C may be written uniquely $f(Y)dX + (g(Y) + Xh(Y))dY$, where f, g and h are polynomials.

Exercise 6.3.8. Let C be circle $x^2 + y^2 = 1$, where $\mathrm{char}(k) \neq 2$.

(a) Show that $D(x)$ and $D(y)$ cover X.
(b) Show that the two fractional differentials $\frac{dx}{y}$ and $-\frac{dy}{x}$ agree on $D(xy)$.
(c) Find a differential ω in $\Omega[X]$ such that $\omega|_{D(y)} = \frac{dx}{y}$ and $\omega|_{D(x)} = -\frac{dy}{x}$. (The correct ω has something to do with the "differential" of area in the case of real curves.)

Exercise 6.3.9. Try to generalize the ideas of Exercise 6.3.8 when $C = \{x^3 + y^3 = 1\}$.

By Lemma 6.3.3 we have an isomorphism $\Omega_X|_x = \mathrm{Cot}_x X$ for all points x of X. This solves the problem of the variation of $\mathrm{Cot}_x X$ with x.

Lemma 6.3.10. *Let X be a variety. Then $\{x \in X | X$ is smooth at $x\}$ is an open dense subset of X.*

Proof. Let x be a point of X at which X is smooth. By Lemma 6.2.3 there is only one component of X passing through x. Thus we may assume that X is irreducible. Then $\{x \in X | X$ is smooth at $x\} = \{x \in X | \dim \Omega_X|_x \leq \dim X\}$ (see Lemma 6.1.2(a)) which is open by Corollary 5.3.4(b). Thus the subset is open. It is dense because it contains the general point on each component. $\square$

Example 6.3.11. Show that the subset of Lemma 6.3.10 is the maximal open subset U such that $\Omega_X|_U$ is locally free.

Next we will make a global calculation of $\Omega_{\mathbf{P}^n}$. Consider the projection $\pi : U = \mathbf{A}^{n+1} - \{0\} \to \mathbf{P}^n$. We have a π-homomorphism $\Omega_{\mathbf{P}^n} \overset{\pi^*}{\to} \Omega_U$. As π is locally a trivial G_m-bundle, π^* identifies $\Omega_{\mathbf{P}^n}$ with a subsheaf of $\pi_* \Omega_{U_{\text{degree zero}}} = \bigoplus_{0 \le i \le n} dX_i \cdot \mathcal{O}_{\mathbf{P}^n}(-1)$.

Lemma 6.3.12. *We have a natural exact sequence*

$$0 \to \Omega_{\mathbf{P}^n} \overset{\pi^*}{\to} \bigoplus_{0 \le i \le n} dX_i \cdot \mathcal{O}_{\mathbf{P}^n}(-1) \overset{\alpha}{\to} \mathcal{O}_{\mathbf{P}^n} \to 0$$

where $\alpha(dX_i \cdot \sigma) = X_i \cdot \sigma$.

Proof. Consider the vector field $R = \sum_{0 \le i \le n} X_i \frac{\partial}{\partial X_i}$ on $\mathbf{A}^{n+1}$. Then R is a radical vector field; i.e. at any point u of U, $R|_u$ points in the direction of the line ku. Thus a differential form ω on U is in $(\pi^* \Omega_{\mathbf{P}^n})$ if and only if ω is degree zero and $< \omega, R > = 0$; i.e. if $\omega = \sum dX_i f_i$ then $\sum X_i f_i = 0$. The lemma follows directly. $\qquad \square$

Define $\omega_{\mathbf{P}^n} = \Lambda^n \Omega_{\mathbf{P}^n}$. Then we have the

Corollary 6.3.13. $\omega_{\mathbf{P}^n} \cong dX_0 \wedge \ldots \wedge dX_n \cdot \mathcal{O}_{\mathbf{P}^n}(-n-1)$.

Thus if $n > 0$ then $\omega_{\mathbf{P}^n}$ and hence $\Omega_{\mathbf{P}^n}$ are not trivial.

Exercise 6.3.14. Show that an algebraic group is smooth at each of its points. (Hint: use translation.)

Proposition 6.3.15. *Let X be a closed subvariety of Y and x be a point of X. If X and Y are smooth let $f_1, \ldots, f_d \in \mathcal{I}_{X,x}$ be elements such that $df_1|_x, \ldots, df_d|_x$ are linearly independent where $d = \dim_x Y - \dim_x X$. Then $\mathcal{I}_X = (f_1, \ldots, f_d)$ in some neighborhood of x. Then the sequence $0 \to \mathcal{I}_X / \mathcal{I}_X^2 \to \Omega_Y|_X \to \Omega_X \to 0$ is an exact sequence of locally free sheaves. Hence if X and Y have pure dimension then $(\det \Omega_Y)|_X \approx \det(\mathcal{I}/\mathcal{I}^2) \otimes \det \Omega_X$.*

If X is a smooth variety of pure dimension n then $\det \Omega_X \equiv \omega_X$ is an invertible sheaf.

Corollary 6.3.16. (Adjunction formula.) *If D is a smooth divisor on X, then $\omega_D \approx \omega_X(D)|_D$.*

Proof. Replace Y by a neighborhood such that the f_i are regular and vanish on X. Let $\mathcal{A} = \mathcal{O}_Y / (f_1, \ldots, f_n)$. Then $G(\mathcal{A}) = \bigoplus_n m_x^n \mathcal{A} / m_x^{n+1} \mathcal{A}$

is generated by the image of $k[\mathrm{Cot}_x(Y)]/(df_1|_x, \ldots, df_n|_x)$ which has dimension $= \dim X$. As $G_x(\mathcal{O}_{X,x})$ is a quotient of GA and is a polynomial ring of dimension $= \dim X$, $G(A) = G_x(\mathcal{O}_{X,x}.)$

As in Lemma 6.2.3 $A_x / \bigcap_j m_x^j A_x$ is an integral domain and it is isomorphic to $\mathcal{O}_{X,x} / \bigcap_j m_x^j$. Now by a generalization of Sublemma 6.2.4, $\bigcap_j m_x^j$ and $\bigcap_j m_x^j A_x$ are zero. Thus $A_x \approx \mathcal{O}_{X,x}$ and $\mathcal{I}_X = (f_1, \ldots, f_d)$ in a neighborhood of x. The other statements are clearly consequences.

$\square$

6.4 Morphisms

Morphisms are quite fascinating. Let $f : X \to Y$ be a morphism. Then for any y in Y we have the closed subvariety $f^{-1}y$ of X. Thus one has the whole family $\{f^{-1}(y)\}$ of varieties attached to a single morphism. In this section we will study how $f^{-1}(y)$ varies with y.

The closure $\overline{f(X)}$ of the image $f(X)$ is the closed subvariety of Y with ideal $\mathcal{I}$ where $\mathcal{I} = \mathrm{Ker}(\mathcal{O}_Y \to f_*\mathcal{O}_X)$.

Proposition 6.4.1. *The image $f(X)$ contains an open dense subset U of its closure $\overline{f(X)}$ such that $\dim f^{-1}(u)$ is locally constant on U. In fact if $\overline{f(X)}$ is irreducible, $\dim f^{-1}(u) = \dim X - \dim \overline{f(X)}$ for u in U.*

Proof. We may assume that $Y = \overline{f(X)}$ and Y is irreducible, and X and Y are affine. The key point is

Lemma 6.4.2. *In this situation we have a non-empty open subset U of Y and that we may factor $f : f^{-1}U \to U$ as $g : f^{-1}U \to U \times \mathbf{A}^n \overset{\pi_U}{\to} U$ where g is a surjective finite morphism.*

First we will note that the lemma implies the proposition. As g and π_U are surjective, $U = f(f^{-1}U)$. Furthermore $\dim f^{-1}(u) = \dim \pi_U^{-1}(u) = \dim u \times \mathbf{A}^n = n$ which is constant. Also $\dim X = \dim f^{-1}U = \dim U + \dim \mathbf{A}^n = \dim U + n$. This shows the proposition. It remains to prove the lemma.

We have an injection $k[Y] \overset{f^*}{\hookrightarrow} k[X]$ where $k[Y]$ is an integral domain and $k[X]$ is a finitely generated algebra. Consider $K = k(Y) \hookrightarrow k[X] \otimes_{k[Y]} k(Y) = A$. We may apply the reasoning of Lemma 1.4.1 to find $K[X_1, \ldots, X_n] \subset A$ such that A is a $K[X_1, \ldots, X_n]$-module of finite type. Let $0 = f_i(y) = y^{j_i} + \cdots + a_{j_i,0}$ be the equations of the integrality of the generators of A over $K[X_1, \ldots, X_n]$. The coefficients $a_{j,i}$ all lie in $k[Y]_{(a)}[X_1, \ldots, X_n]$ for some a in $k[Y]$. Thus $k[Y]_{(a)}[X_1, \ldots, X_n] \subset k[X]_{(f*a)}$ and $k[X]_{(f*a)}$ is a finite type module over the subring. Let $U = D(a) \subset Y$. Then the inclusion gives the required factorizaton. $\qquad\square$

Corollary 6.4.3. *The image of a morphism is the finite union of locally closed subsets.*

Proof. Let $f : X \to Y$ be the morphism. Find an open subset U of $\overline{f(X)}$ as above. Then U is locally closed and $f(X) = U \coprod f(X - f^{-1}U)$. By induction on X this last subset is the union of a finite number of locally closed subsets. $\qquad\square$

We may give an application:

Corollary 6.4.4. *Let $f : H \to G$ be a homomorphism of algebraic groups. Then the image of f is closed.*

Proof. $f(H)$ is a subgroup which contains an open and dense subset U of its closure $\overline{f(H)} = K$. As $f(H) = f(H) \cdot U \supseteq U$, $f(H)$ is open in its closure. Let k be an element of K. Then $kf(H)$ is open and dense in its closure K. Thus $kf(H) \cap f(H)$ is non-empty. Hence $k \in f(H)f(H)^{-1} = f(H)$. Thus $f(H)$ is closed. $\qquad\square$

We will prove

Proposition 6.4.5. *Let $f : X \to Y$ be a morphism. The function $\dim_x f^{-1}(f(x))$ is upper-semicontinuous. If X is irreducible each component of $f^{-1}(f(x))$ has dimension $\geq \dim X - \dim \overline{f(X)}$.*

Proof. We first prove the last statement. We may assume that $Y = \overline{f(X)}$ is affine. Then by Lemma 2.6.8 there exist $\dim Y$ number of regular functions $g_1, \ldots$ such that $f(x)$ is a component of zeroes of $(g_1, \ldots,)$. Thus a component of $f^{-1}(f(x))$ has dimension $\geq \dim X - \dim Y$. Thus the second statement is true.

For the first statement we may assume that X is irreducible. Let U be as in Lemma 6.4.1. Then $\dim f^{-1}(f(x)) = \dim X - \dim Y$ if $f(x) \in U$. Thus $\dim_x f^{-1}(f(x))$ is constant on the open subset $f^{-1}U$ and $\dim_x f^{-1}(f(x)) \geq$ constant for all x. Thus it is enough to prove the proposition for $X = X - f^{-1}(U)$ which we may assume to have been done by induction.   □

6.5 The construction of affine morphisms and normalization

Consider a sheaf $\mathcal{A}$ of $\mathcal{O}_Y$-algebras on a variety Y. We want to know when $\mathcal{A}$ has the form $f_*\mathcal{O}_X$ where $f : X \to Y$ is an affine morphism. We clearly have the necessary conditions, (1) $\mathcal{A}$ is quasi-coherent and (2) for an affine open subvariety V of Y the k-algebra $\mathcal{A}(V)$ is finitely generated and has no nilpotent.

Lemma 6.5.1. *If $\mathcal{A}$ satisfies the conditions (1) and (2) there is a variety $X = \mathrm{Spec}\,\mathcal{A}$ with a morphism $f : X \to Y$ such that $f_*\mathcal{O}_X = \mathcal{A}$. Furthermore if $g : Z \to Y$ is a fixed morphism where Z is a space with functions we have a bijection between the commutative diagram*

$$Z \xrightarrow{h} X$$
$$g \searrow \quad \swarrow f$$
$$Y$$

where h is a morphism, and $\mathcal{O}_Y$-Alg-Hom$(\mathcal{A}, g_*\mathcal{O}_Z)$ which sends h to $h^* : f_*\mathcal{O}_X \to g_*\mathcal{O}_Z$.

Proof. Contrary to the complicated statement the lemma is easy. If V is an open affine let $f^{-1}V$ be $\mathrm{Spec}\,\mathcal{A}(V)$ and the morphism $f : f^{-1}V \to V$ correspond to the k-homomorphism $k[V] \to \mathcal{A}(V)$. To do the construction for a bigger open subset of X, assume that we have constructed $f^{-1}(V_1)$ and $f^{-1}(V_2)$; then to structure $f^{-1}(V_1 \cup V_2)$ one pastes together $f^{-1}(V_1)$ and $f^{-1}(V_2)$ along their open subset $f^{-1}(V_1 \cap V_2)$. The verification is elementary.   □

Let Y be an irreducible variety. Let $L \supset k(Y)$ be a finite field extension of the rational functions on Y. We want to construct an irreducible variety X and a finite surjective morphism $f : X \to Y$ such that $f^* : k(Y) \hookrightarrow k(X)$ is the given field extension.

As this construction problem is stated it does not have a unique solution upto isomorphism. To make the solution unique we require that X is *normal*; i.e. the local rings $\mathcal{O}_{X,x}$ are integrally closed in their quotient field L. The solution to this problem is called the *normalization* of X in L.

Let $\tilde{L}$ be the sheaf on Y such that $\tilde{L}(V) = L$ for all non-empty open subsets V of Y with the obvious restriction. Then $\mathcal{O}_Y \subset \mathbf{Rat}(Y) \subset \tilde{L}$. We need to construct an $\mathcal{O}_Y$-algebra $\mathcal{A} \subset \tilde{L}$. By definition $\mathcal{A}(V) = \bigcap_{v \in V}$ (integral closure in L of $\mathcal{O}_{Y,v}$).

Lemma 6.5.2.

(a) $\mathcal{A}$ is a coherent $\mathcal{O}_Y$-module.

(b) $\mathrm{Spec}(\mathcal{A}) \to Y$ is the normalization of Y in L.

Proof. The problem is local on Y. So we may assume that Y is affine. We have to prove

(1) $\mathcal{A}(Y) =$ integral closure in L of $k[Y]$.

(2) $\mathcal{A}(Y)$ is $k[Y]$-module of finite type.

(3) $X = \mathrm{Spec}(\mathcal{A}(Y))$ is normal and $k(X) = L$.

(4) $\mathcal{A}(D(g)) = \mathcal{A}(Y)_{(g)}$ for all g in $k[Y]$.

If we prove (1) then (2) will be a well-known theorem in algebra. We will just indicate the other steps. Let $B(Y)$ be the right side of (1). As integral closure commutes with localization we have $\mathcal{A}(V) = \bigcap_{v \in V} B(Y)_{n_v}$ but this last group is $\Gamma(Y, \tilde{B}) = B$. Thus (1) holds. (4) is then just the commuting of the two operations. For (3) $\mathcal{A}(Y)$ is integrally closed in its quotient field L. Thus for any point x of X, $\mathcal{A}(Y)_{m_x} = \mathcal{O}_{X,x}$ is integrally closed in $k(X) = L$. $\qquad\square$

We have

Corollary 6.5.3.

Let X be an irreducible variety. Then the subset $U = \{x \in X | \mathcal{O}_{X,x}$ is integrally closed\} is an open dense subset of X.

Proof. Let $f : X' \to X$ be the normalization of X in $k(X)$. U is the complement of the support of the coherent sheaf $f_* \mathcal{O}_{X'} / \mathcal{O}_X$. $\qquad\square$

6.6 Bertini's theorem

Let $X \subset \mathbf{P}^n$ be a closed smooth irreducible subvariety. The set of hyperplanes in $\mathbf{P}^n$ is a projective space $\mathbf{P}^{n*}$. The homogeneous coordinate of a hypersurface $\sum a_i x_i = 0$ in $\mathbf{P}^n = \{x\}$ is $(a_0, \ldots, a_n)$ in $\mathbf{P}^{n*}$. Thus the locus $H = \sum a_i x_i = 0$ in $\mathbf{P}^n \times \mathbf{P}^{n*}$ is a universal family of hypersurfaces.

We want to prove

Theorem 6.6.1. (Bertini's.) *For a in an open dense subset of $\mathbf{P}^{n^*}$ the hyperplane section $X \cap \sum a_i x_i = 0$ is a smooth of dimension $\dim X - 1$.*

Remark. If $\dim X \geq 2$ then the general hyperplane section of X is irreducible.

Proof. Consider the variety Z in $X \times \mathbf{P}^{n^*}$ which consists of a pair (x, a) such that $\sum a_i x_i = 0$ and $x \in X$ and $d(\sum a_i x_i)|_x = 0$. Thus x is a singular point of the hypersurface. The main point is

Claim. $\dim Z = n - 1$.

If we prove the claim then $\pi_{\mathbf{P}^{n^*}} Z$ is a proper closed subset of $\mathbf{P}^{n^*}$. Clearly its complement is the open subset of the theorem.

For the claim consider the projection $\pi : Z \to X$. By dimension theory it will be enough to show that for any x in X the fiber $\pi^{-1}(x)$ has dimension $n - \dim X - 1$. Now the fiber $\pi^{-1}(x)$ is the a such that

$$(*) \qquad \sum a_i x_i = 0 \text{ and } d\left(\sum a_i X_i\right)\big|_x = 0.$$

As $d(\sum a_i x_i)|_x$ is an arbitrary vector in $T_x X$, the equation $(*)$ is $1 + \dim T_x X$ linearly independent equations but as X is smooth, $\dim T_x X = \dim X$. Then $\pi^{-1}(x)$ has the required dimension. $\qquad \square$

7

Curves

7.1 Introduction to curves

A *curve* is an irreducible separated one dimensional variety. A convenient way to use this assumption is that a point c of curve C is determined by the local ring $\mathcal{O}_{C,c}$ in $k(C)$. This is a general fact.

Lemma 7.1.1. *Let x_1 and x_2 be two points on an irreducible separated variety X. If $\mathcal{O}_{X,x_1} \subseteq \mathcal{O}_{X,x_2}$ and $m_{x_1} \subseteq m_{x_2}$ then $x_1 = x_2$.*

Proof. Let V_1 be an affine open neighborhood of x_1. Let $f_1, \ldots, f_n$ be generators of $k[V_1]$. Then these functions are regular at x_1 and, hence, at x_2. Let V_2 be an affine open neighborhood of x_2 where they are regular. Thus $k[V_1] \subseteq k[V_2]$. Then the intersection $V_1 \cap V_2$ is the affine $\mathrm{Spec}(k[V_1] \cdot k[V_2]) = \mathrm{Spec}(k[V_2])$. Thus $V_1 \cap V_2 = V_2$ and $V_2 \subset V_1$. So x_1 and x_2 are in V_1. Now if f is a regular function on V_1 which vanishes at x_1 then $f \in m_{x_1}$ and hence $f \in m_{x_2}$. Thus f vanishes at x_2. So x_2 is in the closure of x_1. Hence $x_2 = x_1$. $\qquad\square$

The local rings of smooth curves are very special. A *discrete valuation ring* is a unique factorization domain with exactly one equivalence class of irreducible elements. Thus if π is irreducible then any non-zero element of the ring may be written uniquely as $(\mathrm{unit})\pi^n$ for $n \geq 0$.

Proposition 7.1.2. *Let C be a curve with a point c. Then the following are equivalent:*

(a) C is smooth at c,
(b) $\mathcal{O}_{C,c}$ is integrally closed, and
(c) $\mathcal{O}_{C,c}$ is a discrete valuation ring.

Proof. $(c) \Rightarrow (b)$ because a UFD is integrally closed (easy algebra). For $(a) \Rightarrow (c)$. We know that m_c/m_c^2 is one dimensional. By Nakayama's lemma if $\pi \in m_c - m_c^2$ then $(\pi) = m_c$. Thus π is irreducible and generates a prime ideal. Let f_0 be a non-zero element of $\mathcal{O}_{C,c}$; we want to show that f_0 has the form (unit)π^n. Otherwise f_0 is infinitely divisible by π. Let $f_n = f/\pi^n$. Then $(f_0) \subseteq (f_1)\ldots$ is an increasing sequence of ideals in the noetherian ring $\mathcal{O}_{C,c}$. Thus $(f_N) = (f_{n+1})$ for $N \gg 0$. So $f/\pi^{N+1} = a\, f/\pi^N$ where a is in $\mathcal{O}_{C,c}$. Hence $1 = a\pi$ which is impossible. Thus any element of $\mathcal{O}_{C,c}$ has the required form. The uniqueness is trivial. This shows $(a) \Rightarrow (c)$.

For $(b) \Rightarrow (a)$, we know that $\mathcal{O}_{C,c}$ is integrally closed. We will try to prove that m_c is principal (hence m_c/m_c^2 is one dimensional). We may assume that C is affine and we have a regular function f on C such that $(f = 0) = \{c\}$. By the nullstellensatz $m_c^N \subset f\,\mathcal{O}_{C,c}$ for some $N \geq 1$.

Claim. Either m_c is principal or $m_c^{N-1} \subset f\,\mathcal{O}_{C,c}$.

It will be enough to prove this claim because if m_c is not principal we will have $\mathcal{O}_{C,c} = m_c^0 \subset f\,\mathcal{O}_{C,c}$ which is impossible.

To prove the claim, let y be an element of m_c^{N-1}. Then $ym_c \subset f\,\mathcal{O}_{C,c}$. Thus $\frac{y}{f}m_c \subset \mathcal{O}_{C,c}$. Either (1) $\frac{y}{f}m_c = \mathcal{O}_{C,c}$ or (2) $\frac{y}{f}m_c \subseteq m_c$ as $\mathcal{O}_{C,c}$ is a local ring. In case (1) $\mathcal{O}_{C,c} = \frac{y}{f}m_c$. Thus $m_c = \frac{f}{y}\mathcal{O}_{C,c}$ is principal. In case (2) happens for all y: We find $\frac{y}{f}$ is integral over $\mathcal{O}_{C,c}$. Hence $\frac{y}{f} \in \mathcal{O}_{C,c}$ is integral and closed. Thus $m_c^{N-1} = \{y\} \subset f\,\mathcal{O}_{C,c}$. This proves the claim. $\square$

In the next section we will need

Lemma 7.1.3.   *Let $f : C \to D$ be a birational morphism between two curves where D is a smooth curve. Then the image of f is open and f gives an isomorphism $C \xrightarrow{\approx} f(C)$.*

Proof. f is an isomorphism between open subsets of C and D. Thus $f(C)$ contains an open subset U of D. Thus the complement of $f(C) \subseteq$ complement of U is finite and hence closed. Thus $f(C)$ is open. Replacing D by $f(C)$ we may assume that f is surjective. We next note that f is injective and $\mathcal{O}_{D,f(c)} = \mathcal{O}_{C,c}$ for all points c

of C. In fact $\mathcal{O}_{C,c} \supset \mathcal{O}_{D,f(c)}$. As D is smooth $\mathcal{O}_{D,f(c)}$ is a DVR, any strictly larger ring $= k(D)$. Thus $\mathcal{O}_{C,c} = \mathcal{O}_{D,f(c)}$ and f is injective by Lemma 7.1.1. It remains to show that f is a homeomorphism. This is clear because f and f^{-1} take finite sets to finite sets. This is enough.

$\square$

7.2 Valuation criterions

We begin with the basic

Lemma 7.2.1. *Let U be an open dense subset of a variety X. Let x be a point of X. Then there is a morphism $f : C \to X$ where C is a smooth curve such that $f^{-1}U$ is not empty and x is contained in the image of f.*

Proof. We may assume that X is irreducible and affine. Let $g : X \to \mathbf{A}^n$ be a finite surjective morphism. As $X - U$ is a proper closed subset of X, $g(X - U)$ is a proper closed subset of $\mathbf{A}^n$. Let V be its complement. Then $g^{-1}(V) \subset U$ by construction. Let L be a straight line connecting $f(x)$ and v where v is a different point of V. (This is always possible when $n \geq 0$); i.e. X is not a point (otherwise there is no problem). Now $g^{-1}(L)$ is a finite cover of L. Thus it is one dimensional. Also L, and hence $g^{-1}(L)$, is defined by $n-1$ equations. Thus by the principal ideal theorem any component of $g^{-1}(L)$ is a curve. Let D be a component passing through x. Then D maps surjectively onto L. Thus $D \cap g^{-1}V \subset D \cap U$ is dense in D. Let $f : C \to D$ be the normalization of D in $k(D)$. Then C is smooth by Proposition 7.1.2. This solves the problem. $\square$

Here is the first criterion.

Proposition 7.2.2. *Let X be a variety. Then X is separated if and only if a morphism $g : U \to X$ from an open dense subset U of a smooth curve C extends in at most one way to a morphism $f : C \to X$.*

Proof.   Assume that X is separated and f_1 and f_2 are two such extensions.   Consider the morphism $\pi : C \to X \times X$ given by $c \mapsto (f_1(c), f_2(c))$. Now $\pi(U) = \Delta_U \subset \Delta_X$ which is closed. Thus $\pi(C) \subset \Delta_X$; i.e. $f_1(c) = f_2(c)$ for all c. Thus $f_1 = f_2$. This proves one way.

Conversely assume that we have uniqueness. Let (x_1, x_2) be a point of $\overline{\Delta_X}$. Then Δ_X is open and dense in $\overline{\Delta_X}$. Thus by Lemma 7.2.1. we may find a morphism $\pi : C \to \overline{\Delta_X}$ such that (x_1, x_2) is in the image and

$\pi^{-1}\Delta_X = U$ is dense in C. Thus $\pi(c) = (f_1(c), f_2(c))$ where $f_1\mid_U = f_2\mid_U$. By uniqueness $f_1 = f_2$. Hence $\pi(C) \subset \Delta_X$. Thus $(x_1, x_2) \in \Delta_X$; i.e. Δ_X is closed.   □

The other criterion is about existence.

Proposition 7.2.3. *Let X be a variety. Then X is complete if and only if all morphisms $g : U \to X$ from a dense open subset U of a smooth curve C extend uniquely to a morphism $f : C \to X$.*

Proof. Assume that X is complete. Let $R = \overline{\text{graph}(g)}$ in $C \times X$. Then R is a curve as X is separated. We want to show that R is graph(f) for some f. To do this we have to show that $\pi_C : R \to C$ is an isomorphism. But π_C is birational. By Lemma 7.1.3 π_C is an isomorphism of R with its image but its image is closed because X is complete. Thus f exists and is clearly unique.

Conversely we will use the statement in the proof of Chow's lemma; i.e. There are an open dense subset U of a projective variety Y and a morphism $h : U \to X$ such that the graph of h is closed in $Y \times X$ and projects onto X. Let y be a point of Y. Let $p : C \to Y$ be a morphism from a smooth curve C such that y is contained in the image of p and $V = p^{-1}U$ is open and dense in C. Consider $g = h \circ p : V \to X$. Let $f : C \to X$ be the extension. Then the image of $(p, f) : C \to Y \times V$ contains the graph of $h\mid_{p(V)}$ as an open dense subset. Thus $\{(p(c), f(c)) \mid c \in C\} \subset \text{graph}(h)$. Thus y is contained in the image of the graph. Hence $U = Y$ and we have a surjective morphism $Y \to X$. Now X is separated by Proposition 7.2.2 and Y is complete because it is projective. Hence X is complete. This settles the case X irreducible and the general case follows.   □

Exercise 7.2.4. How can you prove that a finite union of locally closed subsets of a variety is closed (open)?

7.3 The construction of all smooth curves

A *function field of dimension one* is a field $k \subset K$ such that there is an element t of $K - k$ such that K is a finite algebraic extension of the purely transcendental field $k(t)$.

We want to prove

Theorem 7.3.1. *Given a function field K of dimension one, there is a smooth projective curve C such that $k(C) = K$. Furthermore, any smooth curve D with $k(D) = K$ is uniquely isomorphic to an open subset of C in a way compatible with the identifications.*

Proof. As a set C will be the set of all DVR $\supset k$ with quotient field K. The topology is as usual; i.e. the proper closed subsets are all of the finite subsets. Let D be a smooth curve. Then we have an injective continuous mapping $i : D \to C$ which sends a point d to the local ring $\mathcal{O}_{D,d}$.

Let E be the normalization of $\mathbf{P}^1 = \{t\}$ in K.

Claim. $i : E \to C$ is a homeomorphism.

We need to show that i is surjective. Let R be a DVR $\supset k$ with quotient field K. Either t or t^{-1} is contained in R. Replace t by t^{-1}, if necessary, so we may assume that $t \in R$. Then $B =$ integral closure of $k[t]$ in K, $B \subset R$. Then $p = B \cap m$ is a prime ideal of B where m is a maximal ideal of R. Clearly $p \neq 0$ otherwise each element of $B - \{0\}$ would be a unit in R and hence $K = R$ which is impossible.

Now $B = k[U]$ where U is the inverse image of $\mathbf{A}^1$ in E. Thus p is the maximal ideal of a point e in $U \subset E$. Then $B_p = \mathcal{O}_{E,e} \subset R$. As E is normal $\mathcal{O}_{E,e}$ is a DVR with quotient field K. Hence $\mathcal{O}_{E,e} = R = i(e)$. This shows the claim.

Now we are in a position to describe the curve structure of C intrinsically without reference to t. Let U be an open subset of C. Let $\mathcal{O}_C(U) = \bigcap_{R \in U} R$. We identify $\mathcal{O}_C(U)$ with a ring of functions by taking $f(c) = f((m_c))$ where m_c is the maximal ideal of the ring c. Thus $f(c)$ is a number as $k \xrightarrow{\approx} c/m_c$ is an isomorphism. Clearly the mapping i is an isomorphism.

We next study the functoriality of our construction. Let $L \supset K$ be a finite field extension. Let D be the curve of DVRs in L/k. We have a morphism $D \to C$ which sends the ring $k \subset d \subset L$ to $d \cap K$. To check that this works we identify D with the normalization of C in $L =$ normalization of $\mathbf{P}^1$ in L as before. Then it is clear and we know $D \to C$ is a finite morphism.

Therefore we know that C is complete because any rational mapping $i : D \to C$ extends uniquely to the morphism above. Now that C is complete it is projective because the Chow covering is isomorphic to C (see Lemma 7.1.3). It remains to show that the complement of $i(D)$ is finite. To do this we may assume that D is affine. Let t be a non-constant regular function on D. Then $C - i(D) \subset \{C \mid c \not\ni t\} = \{c \in C \mid t^{-1} \in c$ and $t^{-1}(c) = 0\}$ which is the set of zeroes of a regular function which is finite. This finishes the proof of Theorem 7.3.1.

By Lemma 3.7.2 we know that $k[C] = k$ for our complete curve C. Next we will check

Lemma 7.3.2. *Let C be a smooth curve with $k[C] \neq k$. Then C is affine.*

Proof. Let $t \in k[C] - k$. Then $k[C]$ is a finite extension of $k(t)$. Let $D =$ normalization of $\mathbb{A}^1 = (t)$ in $k[C]$. Then $i : C \hookrightarrow D$ is an isomorphism with an open subset and D is affine. To see that C is affine it is enough to show that i is locally an affine morphism but this is clear. $\square$

Remark. This result is also true for non-smooth curves.

7.4 Coherent sheaves on smooth curves

Let C be a fixed smooth curve. The irreducible divisors on C are just points. Let $D = \sum n_i d_i$ be a divisor. The *degree* of D is $\sum n_i$. Recall that D is effective if each $n_i \geq 0$. Thus the degree of effective divisors is ≥ 0. Recall the theorem that all the local rings $\mathcal{O}_{C,c}$ are UFDs. Thus we may use the theorem which gives an isomorphism $\mathrm{Div}(c) \to \mathrm{IFI}(C)$ given by $D \to \mathcal{O}_C(D)$. This construction gives all invertible sheaves by the discussion in Section 5.5. In fact this gives all non-zero fractional ideals.

Lemma 7.4.1. *Let $\mathcal{I}$ be a non-zero sheaf of fractional ideals on C. Then $\mathcal{I}$ is invertible.*

Proof. Let c be any point of C. Then $\mathcal{I}_c \subset k(C)$ is a non-zero finitely generated $\mathcal{O}_{C,c}$ module. As $\mathcal{O}_{C,c}$ is a DVR, $\mathcal{I}_c = \pi^n \mathcal{O}_{C,c}$ for some n where π is an irreducible element of $\mathcal{O}_{C,c}$. Thus $\mathcal{I}_c$ is free of rank one. Hence $\mathcal{I}$ is locally free of rank one. $\square$

Let $\mathcal{F}$ be an arbitrary coherent sheaf. Let $\mathrm{Rat}(\mathcal{F}) = \varinjlim_{V \neq 0} \mathcal{F}(U)$ be the rational sections of $\mathcal{F}$. Let $\mathbf{Rat}(\mathcal{F})$ be the sheaf on C which assigns $\mathrm{Rat}(\mathcal{F})$ to any non-zero open subset on C with obvious restrictions. If $\mathcal{F} = \mathcal{O}_C$, then $\mathrm{Rat}(\mathcal{O}_C) = k(X)$ and $\mathbf{Rat}(\mathcal{O}_C)$ is the sheaf defined before. As $\mathcal{F}$ is coherent $\mathrm{Rat}(\mathcal{F})$ is a finite dimensional vector space over $k(C)$. We define $\mathrm{rank}(\mathcal{F}) = \dim_{k(C)}\mathrm{Rat}(\mathcal{F})$.

Now $\mathbf{Rat}(\mathcal{F})$ is a quasi-coherent $\mathcal{O}_C$-module and we have a natural $\mathcal{O}_C$-linear mapping $i : \mathcal{F} \to \mathbf{Rat}(\mathcal{F})$. The kernel of $\mathcal{F}$ is a coherent sheaf $\mathcal{F}_{\mathrm{torsion}}$ which consists of the *torsion* in $\mathcal{F}$, where a local section σ of $\mathcal{F}$ is torsional if there exists a non-zero regular local function f such that $f\sigma = 0$. The image $\mathcal{F}'$ of i is *torsion-free*. Thus we have an exact sequence

$$(*) \qquad\qquad 0 \to \mathcal{F}_{\mathrm{torsion}} \to \mathcal{F} \to \mathcal{F}' \to 0.$$

Thus the study of an arbitrary coherent sheaf $\mathcal{F}$ has two parts, the cases where $\mathcal{F}$ is torsional and torsion-free.

If $\mathcal{F}$ is torsional its support consists of a finite number of points $c_1,\ldots,c_d$. Then the stalks $\mathcal{F}_{c_i}$ are finitely generated torsion $\mathcal{O}_{C,c_i}$-modules. Therefore they are finite dimensional k-vector-spaces. We may define an effective divisor $\mathrm{Div}(\mathcal{F}) = \sum_i (\dim_k \mathcal{F}_{c_i}) \cdot c_i$.

Lemma 7.4.2. *Let $\mathcal{F}$ be a torsional coherent sheaf*

(a)   *For any open subset U of C,*

$$\Gamma(U,\mathcal{F}) = \bigoplus_{c_i \in U} \mathcal{F}_{c_i}$$

(b) $\dim_k \Gamma(C,\mathcal{F}) = \deg \mathrm{Div}(\mathcal{F})$.

(c) $\mathcal{F}$ *is flabby.*

Proof.   The statements (b) and (c) follow from (a). For (a) we have an evaluation mapping $\Gamma(U,\mathcal{F}) \to \bigoplus_{c_i \in U} \mathcal{F}_{c_i}$ given by taking germs. It is an isomorphism because $\mathcal{F}|_{\text{complement of support of } \mathcal{F}} = 0$. Thus to show injectivity note that a section of $\mathcal{F}$ is determined by its germs at all points. To show surjectivity an element of $\bigoplus_{c_i \in U} \mathcal{F}_{c_i}$ is the germ of a section σ of $\mathcal{F}$ over a neighborhood in U on support of $\mathcal{F}$. Then extend σ by zero to get it to define over U. $\square$

Next we consider the torsion-free case.

Lemma 7.4.3. *Let $\mathcal{F}$ be a torsion-free coherent sheaf.*

(a) $\mathcal{F}$ *is locally free of rank = $\mathrm{rank}(\mathcal{F})$.*

(b)   *We have a filtration $0 \subset \mathcal{F}_1 \subset \mathcal{F}_2 \cdots \subset \mathcal{F}_{\mathrm{rank}(\mathcal{F})} = \mathcal{F}$ by coherent sheaves $\mathcal{F}_i$ such that all composition factors are invertible.*

Proof. It suffices to show the statement (b). If rank $(\mathcal{F}) = 0$ then $\mathcal{F} = 0$ and there is no problem. If rank$(\mathcal{F}) \geq 0$, let L be a line in Rat$(\mathcal{F})$. Let $\mathcal{F}_1 = \tilde{L} \cap \mathcal{F}$ where $\tilde{L}$ is the constant subsheaf of **Rat**$(\mathcal{F})$ associated to L. Thus $\mathcal{F}_1$ is rank one and isomorphic to a sheaf of fractional ideals if we choose a generator of L. Thus $\mathcal{F}_1$ is invertible and $\mathcal{F}/\mathcal{F}_1 \subset$**Rat**$(\mathcal{F})/\tilde{L}$ is torsion-free of one less rank. Hence we are done by induction. $\square$

Exercise 7.4.4. Show that the sequence $(*)$ splits (unnaturally).

Let $\mathcal{F}$ be a torsion-free coherent sheaf on C. Let D be a divisor. We denote the subsheaf $\mathcal{F} \cdot \mathcal{O}_C(D)$ in **Rat**$(\mathcal{F})$ by $\mathcal{F}(D)$. Here $\mathcal{F}(D)$ is isomorphic to $\mathcal{F} \otimes_{\mathcal{O}_C} \mathcal{O}_C(D)$. If D is effective, the sheaf of rings $\mathcal{O}_C|_D$ is denoted by $\mathcal{O}_D$ and $\mathcal{F}|_D \equiv \mathcal{F}/\mathcal{F}(-D)$. In this situation we have

Lemma 7.4.5.

(a) $\mathcal{F}|_D$ *is torsional.*
(b) Div$(\mathcal{F}|_D) = ($rank $\mathcal{F})D$.
(c) $\dim_k \Gamma(C, \mathcal{F}|_D) = ($rank $\mathcal{F})($deg $D)$.

Proof. Clearly $\mathcal{F}|_D$ is torsional. By Lemma 7.4.2 for (b) and (c) we just have to compute the stalk of $\mathcal{F}|_D$ at a point c. So we may assume that $\mathcal{F}$ is free and the result is trivial. $\square$

7.5 Morphisms between smooth complete curves

Let $f : C \to D$ be a non-constant morphism between smooth complete curves. As $f(C) = D$ we have an inclusion $k(D) \subset k(C)$ which is a finite field extension. Let deg $f \equiv \dim_{k(D)} k(C)$ be the degree of the field extension.

Lemma 7.5.1.

(a) f *is a finite morphism,*
(b) $f_*\mathcal{O}_C$ *is a locally free $\mathcal{O}_D$-module of rank* deg f, *and*
(c) *if $\mathcal{F}$ is a locally free coherent sheaf on C, $f_*\mathcal{F}$ is a locally free $\mathcal{O}_D$-module of rank* $= ($rank $\mathcal{F})($deg $f)$.

Proof. By the construction of complete curves C is the normalization of D in $k(C)$. Thus (a) is true. Now (b) is a special case of (c). To prove (c) note that $f_*\mathcal{F}$ is torsion-free and Rat$(\mathcal{F}) =$Rat$(f_*\mathcal{F})$. Also $f_*\mathcal{F}$ is coherent because $\mathcal{F}$ is. This is a local fact. If U is open and affine in D, then $f^{-1}U$ is affine and $f_*\mathcal{F}(U) = \mathcal{F}(f^{-1}(U))$ which is a finitely generated $k[f^{-1}U]$-module and hence a finite generated $k[U]$-module by (a). Now rank $f_*\mathcal{F} = \dim_{k(D)}\text{Rat}(f_*\mathcal{F}) = \dim_{k(C)}\text{Rat}(\mathcal{F}) \cdot \dim_{k(D)} k(C) = (\text{rank } \mathcal{F})(\deg f)$. $\qquad\square$

Let E be a divisor on D. We may define a divisor $f^{-1}E$ on C by the formula $\mathcal{O}_C(f^{-1}E) = f^*(\mathcal{O}_D(E))$ where $f^*(\mathcal{O}_D(E))$ is considered as a sheaf of fractional ideals on C. Clearly f^{-1} is a homomorphism $\text{Div}(D) \to \text{Div}(C)$.

Lemma 7.5.2. $\deg(f^{-1}E) = \deg(E) \cdot \deg f$.

Proof. We may assume that E is a point d as both sides are homomorphisms. Then $\deg(f^{-1}d) = \dim_k \Gamma(C, f^*\mathcal{O}_d) = \dim_k \Gamma(D, f_*(f^*\mathcal{O}_d)) = \dim_k \Gamma(d, (f_*\mathcal{O}_C)|_d) = \text{rank } f_*\mathcal{O}_C = (\deg f)\cdot 1 = (\deg f)(\deg d)$. $\qquad\square$

Let f be a non-zero rational function on a smooth complete curve C. Then we may define the divisor $\text{div}(f)$ by the formula $f \cdot \mathcal{O}_C = \mathcal{O}_C(-\text{div } f)$. We have a morphism $\tilde{f} : C \to \mathbf{P}^1$ given by $\tilde{f}(c) = f(c)$ when f is regular. This corresponds to the field extension $k(t) \subset k(C)$. This is related to $\text{div}(f)$ by

Lemma 7.5.3. $\text{div}(f) = \tilde{f}^{-1}(0-\infty)$.

Proof. The divisor on $\mathbf{P}^1$, $(\text{div } t)$, where t is the coordinate function, is $0-\infty$. Clearly $\tilde{f}^{-1}(\text{div}(t)) = \text{div}(f)$. So the result follows. $\qquad\square$

This formula has a fantastic global statement.

Corollary 7.5.4. $\deg (\text{div}(f)) = 0$.

Proof. $\deg(\text{div}(f)) = (\deg \tilde{f})(\deg(0-\infty)) = 0$. $\qquad\square$

This important result allows us to define the *degree* of an invertible sheaf $\mathcal{L}$ by the following equation

$$\deg \mathcal{L} = \deg D$$

if $\mathcal{L} \approx \mathcal{O}_C(D)$. The point is that this is well-defined because if D_1 and D_2 are two choices then $D_1 = D_2 + \text{div}(f)$ for some f. Thus $\deg D_1 = \deg D_2$.

We can now make another global statement.

Lemma 7.5.5.

(a) $\dim_k \Gamma(C, \mathcal{L}) \leq \deg \mathcal{L} + 1$.
(b) *In particular* $\Gamma(C, \mathcal{L}) = 0$ *if* $\deg \mathcal{L} < 0$.
(c) *If* $\deg \mathcal{L} = 0$ *then* $\Gamma(C, \mathcal{L}) \neq 0$ *if and only if* $\mathcal{L} \approx \mathcal{O}_C$.

Proof. Let us assume that $\Gamma(C, \mathcal{L}) \neq 0$. Then $\mathcal{L} \approx \mathcal{O}_C(D)$ where D is effective. For (c) if $\deg \mathcal{L} = 0$, then $\deg D = 0$ and hence $D = 0$. Thus (c) is true. For (b) note that $\deg(D) \geq 0$. If $D = 0$, $\Gamma(C, \mathcal{O}_C(D)) = \Gamma(C, \mathcal{O}_C) = k$. Thus (a) is true in this case. If $\deg D > 0$, then $D = d + D_1$ where D_1 is effective. Then we have an exact sequence

$$0 \to \mathcal{O}_C(D_1) \to \mathcal{O}_C(D) \to \mathcal{O}_C(D)|_d \to 0.$$

Take a global section to have

$$0 \to \Gamma(\mathcal{O}_C(D_1)) \to \Gamma(\mathcal{O}_C(D)) \to \Gamma(\mathcal{O}_C(D)|_d).$$

Thus $\dim_k \Gamma(\mathcal{O}_C(D)) \leq \dim_k \Gamma(\mathcal{O}_C(D_1)) + \dim_k \Gamma(\mathcal{O}_C(D)|_d)$ but the last number is one. So we are done by induction. $\qquad\square$

Exercise 7.5.6. Let f be a rational function with only one pole of order 1. Then $\tilde{f} : C \to \mathbf{P}^1$ is an isomorphism.

7.6 Special morphisms between curves

Let $f : C \to D$ be a non-constant morphism between smooth complete curves. Let d be a point of D. Consider the divisor $f^{-1}d = \sum_{c \in f^{-1}d} e_c \cdot c$ where e_c is a positive integer called the *ramification index* of c. If $e_c > 1$ then f is said to be *ramified* at c.

As f corresponds to a finite field extension $k(D) \subset k(C)$. We say that f is *separable (purely inseparable)* if $k(C)/k(D)$ is separable (purely inseparable). By elementary field theory we may factor $f = f_2 \cdot f_1$ where f_2 is separable and f_1 is purely inseparable. Thus the study of a general morphism is reduced to the separable case and the purely inseparable which only happens when the characteristic of k is a prime.

First consider the case when f is separable. Let Tr: $k(C) \to k(D)$ be the trace. Then $\mathrm{Tr}(h \cdot g)$ is a non-degenerate symmetric $k(D)$-bilinear form on $k(C)$. The first fact is that Tr induces an $\mathcal{O}_C$-linear mapping Tr: $f_*\mathcal{O}_C \to \mathcal{O}_D$ because if d is a point of D, we may compute Tr on $f_*\mathcal{O}_{C,d}$ in terms of a free $\mathcal{O}_{D,d}$-module basis of $f_*\mathcal{O}_{C,d}$. Clearly $\mathrm{Tr}|_d : f_*\mathcal{O}_C|_d \to \mathcal{O}_D|_d = k$ is just the trace on the k-algebra $f_*\mathcal{O}_C|_d$.

Consider the mapping $B : f_*\mathcal{O}_C \otimes_{\mathcal{O}_D} f_*\mathcal{O}_C \to \mathcal{O}_D$ induced by $\mathrm{Tr}(f_1 \cdot$

f_2). Then there is an open dense subset U of C such that $B|_U$ is non-degenerate; i.e., $B|_U$ defines an isomorphism $f_*\mathcal{O}_C|_U \to (f_*\mathcal{O}_C|_U)^\wedge$. Let d be a point of U. Then $\mathrm{Tr}|_d : f_*\mathcal{O}_C|_d \otimes f_*\mathcal{O}_C|_d \to k$ is non-degenerate. This means that $\mathrm{Tr}(h_1 \cdot h_2) \neq 0$ for given $h_1 \neq 0$ in $f_*\mathcal{O}_C|_d$ for some h_2. Therefore the ring $f_*\mathcal{O}_C|_d$ has no nilpotents but $f_*\mathcal{O}_C|_d = \mathcal{O}_C|_{f^{-1}d} = \bigoplus_{c \in f^{-1}d} \mathcal{O}_C|_{e_c c}$. Therefore each $e_c = 1$. Therefore f is *unramified* over U.

Exercise 7.6.1. If $h \in f_*\mathcal{O}_{C,d}$, then $\mathrm{Tr}(f)(d) = \sum_{c \in f^{-1}(d)} f(c)$ when d is in U.

We can use calculus to study our separable morphism f. By the previous discussion we know that if π is a parameter at a point d of U then $f^*\pi$ is a parameter at each point of $f^{-1}d$. Thus $f^*(d\pi) = d(f^*\pi)$ is non-zero. Therefore $f^* : \Omega_D \to \Omega_C$ is non-zero. Consider the induced exact sequence $0 \to f^*\Omega_D \to \Omega_C \to \Omega_{C/D} \to 0$ which defines an $\mathcal{O}_C$-module $\Omega_{C/D}$. By construction $\Omega_{C/D}$ is torsional as $\Omega_{C/D}|_{f^{-1}U} = 0$.

Lemma 7.6.2.

(a) We have an isomorphism $f^*\Omega_D(\mathrm{div}\ \Omega_{C/D}) \cong \Omega_C$.

(b) Let $\mathrm{div}\ \Omega_{C/D} = \sum n_c \cdot c$. Then $n_c \geq e_c - 1$ with equality if char $k \backslash e_c$.

Proof. (a) is obvious. For (b) let c be a point of C. Let π be a parameter at $f(c)$. Then $f^*\pi = v \cdot \sigma^{e_c}$ where σ is a parameter at c and v is a unit at c. Thus $f^*d\pi = \sigma^{e_c}dv + e_c\sigma^{e_c-1}vd\sigma$. Thus $f^*d\pi$ vanishes at c to order at least $e_c - 1$ and has exactly this order if char $k\backslash e_c$. On the other hand this order $= n_c$. $\qquad\qquad\square$

If char $k\backslash e_c$ then c is called a *tame* (good!) *ramification* point of f.

Next we consider the purely inseparable case. Let p be the characteristic. Let $q = p^n$ for some $n \geq 1$. Given D we can construct a morphism $F_q : D \to D$ which is purely inseparable of degree q. Consider the q-th power homomorphism $k(D) \hookrightarrow k(D)$. We may identify this extension with $k(D) \hookrightarrow k(D)^{1/q}$. Let $F_q : D_q \to D$ be the normalization of D in $k(D)^{1/q}$. Then we have

Lemma 7.6.3.

(a) F_q has degree q.

(b) F_q is a morphism and the sheaf of rings $F_{q*}(\mathcal{O}_{D_q})$ is isomorphic to $\mathcal{O}_D$ by the q-power mapping.

(c) Any purely inseparable morphism $C \to D$ of degree q is isomorphic
     to F_q.

Proof. We begin with (b). Let d be a point of D and d' be a point of D_q such that $F_q(d') = d$. Then $\mathcal{O}_{D,d} = k(D) \cap \mathcal{O}_{D_q,d'}$. As $\mathcal{O}_{d_q,d'}$ is integrally closed $\mathcal{O}_{D,d}^{1/q} \subset \mathcal{O}_{D_q,d'}$ but $\mathcal{O}_{D,d}^{1/q}$ is a maximal proper subring of $k(D)^{1/q}$. So $\mathcal{O}_{D_q,d'} = D_{D,d}^{1/q}$. As d' is determined by its local ring, d' is uniquely determined by d. Therefore F_q is bijective and takes closed subsets to closed subsets. Hence F_q is a homeomorphism. Also we have shown that $F_{q*}(\mathcal{O}_{D_q}) = \mathcal{O}_{D_q}^{1/q}$. This proves (b). For (a) let π be a parameter at d; then $\pi^{1/q}$ is a parameter at d' on D_q. Thus $F_q^{-1}(d) = qd'$ and hence F_q has degree q.

For (c) $k(D) \subset k(C$ is purely inseparable of degree q. Thus $k(C)^q \subset k(D)$. So $k(D)^{1/q} \supset k(C)$ and they have the same degree over $k(D)$. Hence they are equal. □

7.7 Principal parts and the Cousin problem

Let $\mathcal{F}$ be a locally free coherent sheaf on a smooth curve E. Let c be a point of C. Then the $\mathcal{O}_{C,c}$-module $\mathrm{Rat}(\mathcal{F})/\mathcal{F}_c$ is called the group of *principal parts* at c of rational sections of $\mathcal{F}$. We will denote it by $\mathrm{Prin}_c(\mathcal{F})$. We want to define a sheaf $\mathbf{Prin}(\mathcal{F})$ of principal parts. For any open subset U of C let $\mathbf{Prin}(\mathcal{F})(U) = \bigoplus_{u \in U} \mathrm{Prin}_u(\mathcal{F})$ where restriction to $V \subset U$ forgets coordinates outside of V.

Lemma 7.7.1. *We have an exact sequence of sheaves*

$$0 \to \mathcal{F} \to \mathbf{Rat}(\mathcal{F}) \overset{\alpha}{\to} \mathbf{Prin}(\mathcal{F}) \to 0$$

where $\mathbf{Rat}(\mathcal{F})$ and $\mathbf{Prin}(\mathcal{F})$ are flabby quasi-coherent sheaves.

Proof.   We need to define the homomorphism $\mathbf{Rat}(\mathcal{F})(U) \overset{\alpha(u)}{\to} \mathbf{Prin}(\mathcal{F})(U)$ where U is an open subset. Let f be a section of $\mathbf{Rat}(\mathcal{F})$ over U. Then $f \in \mathrm{Rat}(\mathcal{F})$. There is an open subset V of U such that $f \in \mathcal{F}(V)$. Now $\alpha(U)(f) \equiv \sum_{u \in U} (f_u \text{ modulo } \mathcal{F}_u)$. This is a finite sum because $f_u \in \mathcal{F}_u$ if u is in V. As $\alpha(U)$ commutes with restriction α is defined. Clearly $\mathrm{Ker}(\alpha) = \mathcal{F}$. To see α is surjective we compute stalks at a point c. Then $\alpha_c : \mathrm{Rat}(\mathcal{F}) \to \mathrm{Rat}(\mathcal{F})/\mathcal{F}_c$ is the quotient homomorphism. Thus α is surjective. The rest is evident. □

Taking global sections of the above exact sequence we get that

$$0 \to \Gamma(C, \mathcal{F}) \to \mathrm{Rat}(\mathcal{F}) \xrightarrow{\alpha} \mathrm{Prin}(\mathcal{F})$$

is exact where $\mathrm{Prin}(\mathcal{F}) = \bigoplus_{c \in C} \mathrm{Prin}_c \mathcal{F}$. The *Cousin problem* asks when α is surjective.

Proposition 7.7.2.

(a) α *is surjective if* C *is affine.*

(b) *If* C *is complete and* $\mathcal{F} = \mathcal{O}_C$ *then* α *is surjective if and only if* C *is isomorphic to* $\mathbf{P}^1$.

Proof. (a) follows from Proposition 5.2.3. For (b) if α is surjective we can find a rational function on C with a single simple pole. Thus $C \approx \mathbf{P}^1$ by Exercise 7.5.6. Conversely if $C = \mathbf{P}^1$ let $p \in \mathrm{Prin}(\mathcal{O}_X)$. By (a) for $\mathbf{A}^1$ we may find a rational function f such that $s = p - \alpha(f)$ is zero except at ∞. It suffices to show that s is in the image of α. Now $s = \sum_{1 \leq i \leq n} a_i t^i$ modulo $\mathcal{O}_{\mathbf{P}^1, \infty}$. So $s = \alpha(\sum_{1 \leq i \leq n} a_i t^i)$. $\square$

$\mathrm{Cok}(\alpha)$ is an important global invariant of the sheaf $\mathcal{F}$. We will see that it is the cohomology group $H^1(C, \mathcal{F})$.

We will call the sheaf $\mathcal{F}$ *ordinary* if α is surjective; i.e., the Cousin problem for $\mathcal{F}$ has a positive solution.

We will need a criterion for the sheaf $\mathcal{F}$ to be ordinary.

Lemma 7.7.3. $\mathcal{F}$ *is ordinary if and only if, for all effective divisors* D *and all points* c *of* C, $\dim_k \Gamma(C, \mathcal{F}(D + c)) / \Gamma(C, \mathcal{F}(D)) = \mathrm{rank}\, \mathcal{F}$.

Proof. Let E be an effective divisor. Then $\mathcal{F}(E)/\mathcal{F}$ is isomorphic to a $(\deg E \times \mathrm{rank}\, \mathcal{F})$ dimensional subspace of $\mathrm{Prin}(\mathcal{F})$. The image of α intersected with this subspace is isomorphic to $\Gamma(C, \mathcal{F}(E))/\Gamma(C, \mathcal{F})$. Thus $\mathcal{F}$ is ordinary iff $\dim \Gamma(C, \mathcal{F}(E))/\Gamma(C, \mathcal{F})) = \deg E \times \mathrm{rank}\, \mathcal{F}$ for all E's, i.e. each time that we add a point c to E $\dim \Gamma(C, \mathcal{F}(E + c))/\Gamma(C, \mathcal{F}(E)) = \mathrm{rank}\, \mathcal{F}$. $\square$

8

Cohomology and the
Riemann–Roch theorem

8.1 The definition of cohomology

Let $\mathcal{F}$ be an abelian sheaf on a topological space X. Recall that we have the exact sequences $0 \to \mathcal{F} \to D(\mathcal{F}) \to D(\mathcal{F})/\mathcal{F} \to 0$. We may iterate this construction as follows: Let $C^0(\mathcal{F}) = \mathcal{F}$, $D^i(\mathcal{F}) = D(C^i(\mathcal{F}))$ and $0 \to C^i(\mathcal{F}) \to D^i(\mathcal{F}) \to C^{i+1}(\mathcal{F}) \to 0$ be the canonical quotient sequence. Putting these short exact sequences together we get a resolution $\mathcal{F} \to D^0(\mathcal{F}) \to D^1(\mathcal{F}) \to \ldots$ of $\mathcal{F}$ by the complex $D^\star(\mathcal{F})$. The sheaves $D^i(\mathcal{F})$ are flabby by construction. And the construction is functorial in $\mathcal{F}$.

We get a complex $\Gamma(X, D^\star\mathcal{F})$ by taking global sections. The i-homology group of this complex is the *i-th cohomology group* $H^i(X, \mathcal{F})$. Clearly $H^i(X, -)$ is an additive functor. We have a natural mapping $\Gamma(X, \mathcal{F}) \to H^0(X, \mathcal{F})$, which is easily seen to be an isomorphism.

A general problem is to compute these cohomology groups because the definition is not very enlightening. This generalizes the problem of computing the space $\Gamma(X, \mathcal{F})$ of global sections of $\mathcal{F}$. In this section we will develop some general methods to approach this calculation.

Let $0 \to \mathcal{F}_1 \to \mathcal{F}_2 \to \mathcal{F}_3 \to 0$ be a short exact sequence of abelian sheaves on X. Then we have an exact sequence of complexes of sheaves

$$0 \to D^\star(\mathcal{F}_1) \to D^\star(\mathcal{F}_2) \to D^\star(\mathcal{F}_3) \to 0.$$

As these sheaves are all flabby we have an exact sequence of complexes of groups

$$0 \to \Gamma(X, D^\star(\mathcal{F}_1)) \to \Gamma(X, D^\star(\mathcal{F}_2)) \to \Gamma(X, D^\star(\mathcal{F}_3)) \to 0.$$

By the usual snake lemma we have a long exact sequence of cohomology groups,

$$0 \to H^0(X, \mathcal{F}_1) \to H^0(X, \mathcal{F}_2) \to H^0(X, \mathcal{F}_3) \overset{\delta}{\to}$$

$$H^1(X, \mathcal{F}_1) \to H^1(X, \mathcal{F}_2) \to H^1(X, \mathcal{F}_3) \overset{\delta}{\to} H^2(X, \mathcal{F}_1) \to \cdots.$$

This is the most important property of cohomology.

Lemma 8.1.1. *If $\mathcal{F}$ is flabby, then $H^i(X, \mathcal{F}) = 0$ for $i > 0$.*

Proof. We know that $D(\mathcal{F})/\mathcal{F}$ is flabby by Lemma 4.3.2. Furthermore, the sequence $0 \to \Gamma(X, \mathcal{F}) \to \Gamma(X, D(\mathcal{F})) \to \Gamma(X, D(\mathcal{F})/\mathcal{F}) \to 0$ is exact. Repeating this idea we see that the whole complex $0 \to \Gamma(X, \mathcal{F}) \to \Gamma(X, D^\star\mathcal{F})$ is exact. Hence $H^i(X, \mathcal{F}) = 0$ for $i > 0$. $\square$

The next result is a general resolution principle.

Lemma 8.1.2. *Let $0 \to \mathcal{F} \to \mathcal{F}^0 \to \mathcal{F}^1 \to \cdots$ be a resolution of $\mathcal{F}$ by sheaves $\mathcal{F}^i$ such that $H^j(X, \mathcal{F}^i) = 0$ for all $j > 0$ and all i. Then $H^i(X, \mathcal{F})$ is naturally isomorphic to the i-homology group of the complex*

$$\Gamma(X, \mathcal{F}^0) \to \Gamma(X, \mathcal{F}^1) \to \cdots.$$

Proof. Consider the short exact sequence

$$0 \to \mathcal{F} \to \mathcal{F}^0 \to \mathcal{G} \to 0$$

and the resolution $0 \to \mathcal{G} \to \mathcal{F}^1 \to \mathcal{F}^2 \ldots$.

By left exactness of $\Gamma(X,-)$ the statement is trivial when $i = 0$. If $i = 1$ then

$$\Gamma(X, \mathcal{F}^0) \to \Gamma(X, \mathcal{G}) \to H^1(X, \mathcal{F}) \to 0$$

is exact and

$$0 \to \Gamma(X, \mathcal{G}) \to \Gamma(X, \mathcal{F}^1) \to \Gamma(X, \mathcal{F}^2)$$

is exact.

Thus $H^1(X, \mathcal{F}) \approx$ first homology group of $\Gamma(X, \mathcal{F}^\star)$. If $i > 1$ then $H^{i-1}(X, \mathcal{G}) \overset{\delta}{\to} H^i(X, \mathcal{G})$ is an isomorphism and by induction $H^{i-1}(X, \mathcal{G})$ is the $(i - 1)$-homology of the complex $\Gamma(X, \mathcal{F}^1) \to \Gamma(X, \mathcal{F}^2) \to \cdots$ which is the i-homology group of the complex $\Gamma(X, \mathcal{F}^\star)$. Thus the lemma follows. $\square$

Lemma 8.1.3.   *Let* $(\mathcal{F}_i)$ *be a direct system of abelian sheaves on a noetherian topological space* X. *Then we have a natural isomorphism*

$$\varinjlim H^j(X, \mathcal{F}_i) \overset{\approx}{\to} H^j(X, \varinjlim \mathcal{F}_i).$$

Proof.   We have the resolution $0 \to \varinjlim \mathcal{F}_i \to \varinjlim D^*(\mathcal{F}_i)$. By Corollary 4.4.4 each $\varinjlim D^j(\mathcal{F}_i)$ is flabby. So by Lemma 8.1.1 the hypothesis of Lemma 8.1.2 is satisfied. Thus $H^j(X, \varinjlim D^*(\mathcal{F}_i)) = j$-homology of $\Gamma(X, \varinjlim D^*(\mathcal{F}_i))$. By Lemma 4.4.3 this is the j-homology of $\varinjlim (\Gamma(X, D^*(\mathcal{F})) = \varinjlim$ of the j-homology of $\Gamma(X, D^*(\mathcal{F}_i)) = \varinjlim H^j(X, \mathcal{F}_i)$. $\square$

8.2 Cohomology of affines

We begin with a local vanishing principle. Recall if U is an open subset of a topological space X, and $\mathcal{F}$ is a sheaf on X, we have defined in Section 5.2. a sheaf $_U\mathcal{F}$ and a mapping $\mathcal{F} \to {}_U\mathcal{F}$. Clearly $_U\mathcal{F}$ is flabby.

Proposition 8.2.1.   *Let* $\mathcal{V}$ *be a basis of open subsets closed under finite intersection of a topological space* X. *Let* $\mathcal{F}$ *be an abelian sheaf on* X. *Assume that* $H^j(V, \mathcal{F})$ *is zero for* $0 < j < i$ *and all* V *in* $\mathcal{V}$. *For any element* σ *in* $H^i(X, \mathcal{F})$, *we may find an open covering* $X = \bigcup W_\alpha$ *by members* W_α *of* $\mathcal{V}$ *such that the image of* σ *in* $H^i(X, {}_{W_\alpha}\mathcal{F})$ *is zero for each* α.

Proof.   Let $i = 1$. We have an exact commutative diagram,

$$
\begin{array}{ccccccccc}
0 & \longrightarrow & \mathcal{F} & \longrightarrow & D(\mathcal{F}) & \longrightarrow & D(\mathcal{F})/\mathcal{F} & \longrightarrow & 0 \\
& & \downarrow & & \downarrow & & \downarrow & & \\
0 & \longrightarrow & {}_W\mathcal{F} & \longrightarrow & {}_WD(\mathcal{F}) & \longrightarrow & {}_WD(\mathcal{F})/{}_W\mathcal{F} & \longrightarrow & 0
\end{array}
$$

The cohomology class σ in $H^1(X, \mathcal{F})$ is represented by a section τ of $D(\mathcal{F})/\mathcal{F}$ over X. As the first row is exact and $\mathcal{V}$ is a basis, we may find a covering $X = \bigcup W_\alpha$ by members W_α of $\mathcal{V}$ such that $\tau|_{W_\alpha}$ lifts to a section of $D(\mathcal{F})$ over W_α. Therefore, for any α, the image of τ in $\Gamma(X, {}_{W_\alpha}(D(\mathcal{F}))/{}_{W_\alpha}\mathcal{F})$ lifts to a section of $_{W_\alpha}D(\mathcal{F})$ over X. Hence, the image of σ in $H^1(X, {}_{W_\alpha}\mathcal{F})$ must be zero.

Assume that $i > 1$. Let W be a member of $\mathcal{F}$. I claim that (*a*)   there is a commutative exact diagram,

$$0 \longrightarrow \mathcal{F} \longrightarrow D(\mathcal{F}) \longrightarrow D(\mathcal{F})/\mathcal{F} \longrightarrow 0$$

$$0 \longrightarrow {}_w\mathcal{F} \longrightarrow {}_wD(\mathcal{F}) \longrightarrow {}_w(D(\mathcal{F})/\mathcal{F}) \longrightarrow 0$$

(*b*)   the sheaf $D(\mathcal{F})/\mathcal{F}$ satisfies the assumption of the lemma with i replaced by $i-1$.

First, I will show how the claim implies the lemma. As $D(\mathcal{F})$ and ${}_wD(\mathcal{F})$ are flabby, we have isomorphisms in the commutative diagram

$$H^{i-1}(X, D(\mathcal{F})/\mathcal{F}) \xrightarrow{\approx} H^i(X, \mathcal{F})$$

$$H^{i-1}(X, {}_w(D(\mathcal{F})/\mathcal{F})) \xrightarrow{\approx} H^i(X, {}_w\mathcal{F})$$

as $i > 1$. Therefore, our proposition follows from the claim by induction on i.

To prove the claim, let V be any member of $\mathcal{V}$. By our assumption and Lemma 4.3.3, we have a short exact sequence,

$$0 \longrightarrow \Gamma(W \cap V, \mathcal{F}) \longrightarrow \Gamma(W \cap V, D(\mathcal{F})) \longrightarrow \Gamma(W \cap V, D(\mathcal{F})/\mathcal{F}) \longrightarrow 0$$

$$\Gamma(V, {}_w\mathcal{F}) \qquad \Gamma(V, {}_wD(\mathcal{F})) \qquad \Gamma(V, {}_w(D(\mathcal{F})/\mathcal{F}))$$

Thus, as $\mathcal{V}$ is a basis and $i > 1$, the bottom line in claim (*a*) is exact (this was the only point in (*a*) at issue). Furthermore, we have isomorphisms, $H^j(V, D(\mathcal{F})/\mathcal{F}) \approx H^{j+1}(V, \mathcal{F})$ for $j > 0$. Hence part (*b*) follows from our assumption on $\mathcal{F}$. $\square$

We now are prepared for

Theorem 8.2.2.   (Serre) *Let $\mathcal{F}$ be a quasi-coherent sheaf on an affine variety X. Then $H^i(X, \mathcal{F}) = 0$ for all $i > 0$.*

Proof.   We will prove the theorem by induction on i. Consider the basis $\{D(f)\}$ of X where f is a regular function. Each member of this basis is affine and closed under finite intersection. Thus we may apply Proposition 8.2.1. Given α in $H^i(X, \mathcal{F})$ we may find a finite open covering $U_1, \ldots, U_d$ by basis element such that α lies in $H^i(X, \mathcal{F}_{U_j})$ for each j.

Consider the exact sequence

$$0 \to \mathcal{F} \to \bigoplus_{U_j} \mathcal{F} \to \mathcal{G} \to 0$$

of quasi-coherent sheaves. Then by the long exact sequence $\alpha = \delta(\beta)$ for some β in $H^{i-1}(X, \mathcal{G})$. If $i > 1$, by induction β is zero and, hence, α is zero. If $i = 1$, $\delta = 0$ because $\Gamma(X, -)$ is exact for quasi-coherent sheaves on an affine. Thus $\alpha = 0$ here also.   □

A useful form of this theorem is

Corollary 8.2.3. *Let $f : X \to Y$ be an affine morphism of varieties. Let $\mathcal{F}$ be a quasi-coherent $\mathcal{O}_X$-module. Then for all i we have a natural isomorphism*

$$H^i(X, \mathcal{F}) \xrightarrow{\approx} H^i(Y, f_\ast \mathcal{F}).$$

Proof. Let $\mathcal{F} \to \mathcal{F}^\ast$ be a flabby resolution of $\mathcal{F}$. Then for all open affine subvarieties V of Y, $0 \to \Gamma(f^{-1}V, \mathcal{F}) \to \Gamma(f^{-1}V, \mathcal{F}^\ast)$ is exact because $f^{-1}V$ is affine. Hence $f_\ast \mathcal{F} \to f_\ast \mathcal{F}^\ast$ is a flabby resolution of $f_\ast \mathcal{F}$. Thus

$$H^i(Y, f_\ast \mathcal{F}) = i\text{-homology group of } \Gamma(Y, f_\ast \mathcal{F}^\ast)$$
$$= i\text{-homology group of } \Gamma(X, \mathcal{F}^\ast)$$
$$= H^i(X, \mathcal{F}).$$

□

Thus affine morphisms don't change cohomology groups.

8.3 Higher direct images

Let $f : X \to Y$ be a continuous mapping of topological spaces. Let $\mathcal{F}$ be an abelian sheaf on X. By definition $R^i f_\ast \mathcal{F}$ is the i-homology sheaf of the complex $f_\ast(D^\ast(\mathcal{F}))$. As $f_\ast$ is left exact, $f_\ast \mathcal{F} \cong R^0 f_\ast \mathcal{F}$. One way to compute $R^i f_\ast \mathcal{F}$ is to note that it is the sheaf associated to the presheaf $V \to H^i(f^{-1}V, \mathcal{F})$ for any open subset V of Y.

If $0 \to \mathcal{F}_1 \to \mathcal{F}_2 \to \mathcal{F}_3 \to 0$ is an exact sequence of abelian sheaves on X, then we have a long exact sequence

$$0 \to R^0 f_\ast \mathcal{F}_1 \to R^0 f_\ast \mathcal{F}_2 \to R^0 f_\ast \mathcal{F}_3 \xrightarrow{\delta}$$

$$R^1 f_\ast \mathcal{F}_1 \to R^1 f_\ast \mathcal{F}_2 \to R^1 f_\ast \mathcal{F}_3 \xrightarrow{\delta} R^2 f_\ast \mathcal{F}_1 \to \cdots,$$

in the same way as cohomology groups. If Y is a point, $R^i f_* \mathcal{F}(\text{point}) = H^i(X, \mathcal{F})$. It is very convenient to have higher direct images.

We may restate Serre's theorem as

Lemma 8.3.1. *If $f : X \to Y$ is an affine morphism of varieties and $\mathcal{F}$ is a quasi-coherent $\mathcal{O}_X$-module then $R^i f_* \mathcal{F} = 0$ for $i > 0$.*

Proof. Same as the proof of the last corollary. $\qquad\square$

We will break some new ground with

Proposition 8.3.2. *Let $f : X \to Y$ be a morphism of varieties and let $\mathcal{F}$ be a quasi-coherent $\mathcal{O}_X$-module. Then for all i*

(a) $R^i f_* \mathcal{F}$ *is a quasi-coherent $\mathcal{O}_Y$-module, and*

(b) *for each affine open subset V of Y, $H^i(\widetilde{f^{-1}V}, \mathcal{F}) \cong (R^i f_* \mathcal{F})|_V$.*

Proof. As (b) implies (a), we will prove (b). We may assume that Y is affine. We have the mapping $H^i(X, \mathcal{F}) \to \Gamma(Y, R^i f_* \mathcal{F})$. Thus we have an $\mathcal{O}_Y$-homomorphism $(H^i(X, \mathcal{F}))^{\widetilde{}} \to R^i f_* \mathcal{F}$. We want to show that this is an isomorphism. We need to check that it is an isomorphism of stalks at any point y of Y; i.e. $H^i(X, \mathcal{F})_{n_y} \to \underrightarrow{\lim} H^i(f^{-1}V, \mathcal{F})$ where $y = V$ open in Y is an isomorphism where n_y is the ideal of y in $k[Y]$. To do this it is enough to see that $H^i(X, \mathcal{F})_{(g)} \to H^i(f^{-1}(D(g)), \mathcal{F})$ is an isomorphism for all g in $k[Y]$. Now the inclusion $f^{-1}(D(g)) = D(f^* g) \subset X$ is affine. Thus $H^i(f^{-1}(D(g)), \mathcal{F}) = H^i(X, {}_{D(f^* g)}\mathcal{F})$ but $\mathcal{F}$ is quasi-coherent. So $\underrightarrow{\lim} \frac{1}{(f^* g)^n} \mathcal{F} = {}_{D(f^* g)}\mathcal{F}$. Thus $H^i(X, {}_{D(f^* g)}\mathcal{F}) = \underrightarrow{\lim} \frac{1}{(f^* g)^n} H^i(X, \mathcal{F}) = H^i(X, \mathcal{F})_{(g)}$ so (b) is true. $\qquad\square$

Thus direct images and cohomology are equivalent languages. We will compute a direct image in the simplest possible situation.

Lemma 8.3.3. *Let X and Y be two varieties. Let $\mathcal{F}$ be a quasi-coherent sheaf on Y. Then for all i we have an isomorphism $\mathcal{O}_X \otimes_k H^i(Y, \mathcal{F}) \xrightarrow{\cong} R^i \pi_{X*}(\pi_Y^* \mathcal{F})$.*

This result says that the variational cohomology of a constant family of sheaves is constant.

Proof. The arrow is just given by multiplication and the pull-back mapping $H^i(Y, \mathcal{F}) \to H^i(X \times Y, \pi_Y^* \mathcal{F}) \to \Gamma(X, R^i \pi_{X*}(\pi_Y^* \mathcal{F}))$. Thus we may check that it is an isomorphism locally on X. So we may assume that X is affine. Then by Proposition 8.3.2 we see that the obvious mapping

$$k[X] \otimes_k H^i(Y, \mathcal{F}) \to H^i(X \times Y, \pi_Y^* \mathcal{F})$$

is an isomorphism. As X is affine, π_Y is an affine morphism. Thus

$$H^i(X \times Y, \pi_Y^* \mathcal{F}) = H^i(Y, \pi_{Y*} \pi_Y^* \mathcal{F})$$

but $\pi_{Y*} \pi_Y^* \mathcal{F} = k[X] \otimes_k \mathcal{F}$. Thus as Y is noetherian $H^i(Y, \pi_{Y*} \pi_Y^* \mathcal{F}) = H^i(Y, k[X] \otimes_k \mathcal{F}) = k[X] \otimes_k H^i(Y, \mathcal{F})$.   □

8.4 Beginning the study of the cohomology of curves

Let $\mathcal{F}$ be a locally free coherent sheaf on a smooth curve C. We have a flabby resolution

$$0 \to \mathcal{F} \to \mathbf{Rat}(\mathcal{F}) \to \mathbf{Prin}(\mathcal{F}) \to 0$$

by Section 7.7. Thus $H^i(C, \mathcal{F}) = 0$ if $i > 1$ and we have an exact sequence

$$0 \to H^0(C, \mathcal{F}) \to \mathbf{Rat}(\mathcal{F}) \to \mathbf{Prin}(\mathcal{F}) \to H^1(C, \mathcal{F}) \to 0.$$

Thus $H^1(C, \mathcal{F})$ measures the obstruction to solving the Cousin problem. Note if $\mathcal{G}$ is a torsional coherent sheaf then $H^i(C, \mathcal{G}) = 0$ if $i > 0$ because $\mathcal{G}$ is flabby. Thus $H^1(C, \mathcal{F}) = 0$ if and only if $\mathcal{F}$ is ordinary.

In this section we will eventually prove

Theorem 8.4.1. *Assume that C is complete.*

(a)   *The cohomology groups $H^0(C, \mathcal{F})$ and $H^1(C, \mathcal{F})$ are finite dimensional k-vector-spaces.*

(b) $\dim_k H^0(C, \mathcal{F}) - \dim_k H^1(C, \mathcal{F}) = \deg(\det \mathcal{F}) + (\operatorname{rank} \mathcal{F})(1 - g)$ *where $g \equiv \dim_k H^1(C, \mathcal{O}_C)$ is the genus of C.*

The expression $\dim_k H^0(C, \mathcal{F}) - \dim_k H^1(C, \mathcal{F})$ is called the *Euler-characteristic* of $\mathcal{F}$ and is denoted by $\chi(\mathcal{F})$.

Step 1.   We reduce to the case where $\mathcal{F}$ is invertible. If $\mathcal{F}$ has rank > 1 then we have an exact sequence $0 \to \mathcal{L} \to \mathcal{F} \to \mathcal{G} \to 0$ where $\mathcal{L}$ is invertible and $\mathcal{G}$ is locally free of rank $=$ rank $\mathcal{F} - 1$. Thus by induction we know the theorem for $\mathcal{L}$ and $\mathcal{G}$. Consider the long exact sequence

$$0 \to H^0(C, \mathcal{L}) \to H^0(C, \mathcal{F}) \to H^0(C, \mathcal{G}) \to$$

$$H^1(C, \mathcal{L}) \to H^1(C, \mathcal{F}) \to H^1(C, \mathcal{G}) \to 0.$$

Thus (a) for $\mathcal{L}$ and $\mathcal{G}$ implies (a) for $\mathcal{F}$ and

$$\chi(\mathcal{F}) = \chi(\mathcal{L}) + \chi(\mathcal{G})$$
$$= \deg \mathcal{L} + (1 - g) + \deg(\det \mathcal{G}) + (\text{rank } \mathcal{F} - 1)(1 - g)$$
$$= \deg(\mathcal{L} \otimes \det \mathcal{G}) + (\text{rank } \mathcal{F})(1 - g)$$
$$= \deg(\det \mathcal{F}) + (\text{rank } \mathcal{F})(1 - g).$$

Hence (b) is true for $\mathcal{F}$ if it is true for $\mathcal{L}$ and $\mathcal{G}$.

Step 2. It suffices to find one invertible sheaf $\mathcal{L}$ such that $H^1(C, \mathcal{L})$ is finite dimensional. Recall that we know that $H^0(C, \mathcal{M})$ is finite dimensional for any invertible sheaf $\mathcal{M}$. Let c be a point of C.

We have an exact sequence

$$0 \to \mathcal{M} \to \mathcal{M}(c) \to \mathcal{M}(c)|_c \to 0.$$

Thus we have a long exact sequence

$$0 \to H^0(C, \mathcal{M}) \to H^0(C, \mathcal{M}(c)) \to H^0(C, \mathcal{M}(c)|_c) \to$$
$$H^1(C, \mathcal{M}) \to H^1(C, \mathcal{M}(c)) \to 0$$

where $H^0(C, \mathcal{M}(c)|_c)$ is one dimensional.

Thus $H^1(C, \mathcal{M}(c))$ is finite dimensional iff $H^1(C, \mathcal{M})$ is and this case $\chi(\mathcal{M}(c)) = \chi(\mathcal{M}) + 1$.

Now any $\mathcal{M}$ has the form $\mathcal{L}(\sum c_i - \sum d_j)$. Thus if $\mathcal{L}$ has finite H^1 then so does $\mathcal{M}$ and $\chi(\mathcal{M}) = \chi(\mathcal{L}) + \deg(\sum c_i - \sum d_j)$. Thus we may assume that $\mathcal{L} = \mathcal{O}_C$. Then we get if $\mathcal{M} = \mathcal{O}_C(D)$,

$$\chi(\mathcal{M}) = \deg(D) + \chi(\mathcal{O}_C)$$
$$= \deg \mathcal{M} + \dim \Gamma(C, \mathcal{O}_C) - \dim_k H^1(C, \mathcal{O}_C)$$
$$= \deg \mathcal{M} + 1 - g.$$

Thus (a) and (b) would be true for formal reasons if there existed one invertible $\mathcal{L}$ with finite H^1.

Consider where $\mathcal{M}$ is a locally free coherent sheaf on C.

Lemma 8.4.2. *If* $\Gamma(C, \Omega_C \otimes \mathcal{M}^\vee) = 0$, $H^1(C, \mathcal{M}) = 0$.

This will solve our problem because if $\deg \mathcal{L} > \deg \Omega_C$ then $H^1(C, \mathcal{L}) = 0$ because $\Gamma(C, \Omega_C \otimes_{\mathcal{O}_C} \mathcal{L}^{\otimes -1}) = 0$ because $\deg \Omega_C \otimes \mathcal{L}^{\otimes -1} < 0$ by Lemma 7.5.5(b).

Corollary 8.4.3. *If* $\mathcal{M}$ *is an invertible sheaf of degree* $> \deg \Omega_C$ *then* $H^1(C, \mathcal{M}) = 0$.

The proof of the lemma is important because the method will eventually prove the Riemann–Roch theorem.

Proof of Lemma 8.4.2. Let $\mathcal{F} = \mathcal{M}(D)$ for some effective divisor D. Then $\Gamma(C, \Omega_C \otimes \mathcal{F}^\wedge) = \Gamma(C, (\Omega_C \otimes \mathcal{M}^\wedge)(-D)) \subset \Gamma(C, \Omega_C \otimes \mathcal{M}^\wedge) = 0$ is zero.

Let c be a point of C. We have the exact sequence $0 \to \mathcal{F} \to \mathcal{F}(c) \to \mathcal{F}(c)|_c \to 0$. Then we have the differential $\delta_c : H^0(C, \mathcal{F}(c)|_c) \to H^1(C, \mathcal{F})$.

Step 1. δ_c is zero for all c.

The idea of the proof is to see how δ_c varieties with c *globally* in C. Consider the exact sequence

$$0 \to \pi_1^* \mathcal{F} \to \pi_1^* \mathcal{F}(\Delta) \to \pi_1^* \mathcal{F}(\Delta)|_\Delta \to 0$$

on the product $C \times C$. Then taking π_{2*} we have a differential

$$\delta : \pi_{2*}(\pi_1^* \mathcal{F}(\Delta)|_\Delta) \to R^1 \pi_{2*} \pi_1^* \mathcal{F} = H^1(\mathcal{F}) \otimes_k \mathcal{O}_C.$$

Clearly the value $\delta(c)$ of δ at c is just δ_c because π_2 is an isomorphism of Δ to C. Thus we want to show that δ is zero. Now $\mathcal{O}_{C \times C}(-\Delta)|_\Delta \approx \Omega_C$. Thus $\pi_{2*}(\pi_1^* \mathcal{F}(\Delta)|_\Delta)$ is just $\mathcal{F} \otimes_{\mathcal{O}_C} \Omega_C^{\otimes -1}$. Thus $\delta : \mathcal{F} \otimes_{\mathcal{O}_C} \Omega_C^{\otimes -1} \to H^1(\mathcal{F}) \otimes_k \mathcal{O}_C$ has dual $\delta^\wedge : H^1(\mathcal{F})^\wedge \otimes_k \mathcal{O}_C \to \Omega_C \otimes \mathcal{F}^\wedge$ which must be zero because $\Omega_C \otimes \mathcal{F}^\wedge$ has no non-zero sections. Thus $\delta = 0$. This settles Step 1.

Step 2. $H^1(C, \mathcal{M}) = 0$.

By Lemma 7.7.3, to show that $\mathcal{F}$ is ordinary we need the following. By Step 1, we have an exact sequence

$$0 \to \Gamma(C, \mathcal{F}(D)) \to \Gamma(C, \mathcal{F}(D)(c)) \to \Gamma(C, \mathcal{F}(D)(c)|_c) \to 0.$$

where D is an effective divisor. Then we are done because

$$\dim_k \Gamma(C, \mathcal{F}(D)(c)|_c) = \operatorname{rank} \mathcal{F}.$$

Corollary 8.4.4.

(a) $\dim_k \Gamma(C, \mathcal{F}) \geq \deg(\det \mathcal{F}) + (\operatorname{rank} F)(1 - g)$

(b) $\dim_k H^1(C, \mathcal{F}) \geq -\deg(\det \mathcal{F}) - (\operatorname{rank} F)(1 - g)$.

Corollary 8.4.5. *Any proper open subset U of C is affine.*

Proof. Let c be point not in U. Then $\dim \Gamma(C, \mathcal{O}_C(nc)) > 1$ if $n \gg 0$. Thus there is a non-constant regular function on $U \subset C - \{c\}$. Hence U is affine by Lemma 7.3.2. □

8.5 The Riemann–Roch theorem

Now we know that $H^1(C, \mathcal{L}) = 0$ if $\deg \mathcal{L} > \deg \Omega_C$ and $H^1(C, \mathcal{L}) \neq 0$ if

$\deg \mathcal{L} \ll 0$. Thus there is an invertible sheaf $\mathcal{M}$ such that $H^1(C,\mathcal{M}) \neq 0$ and $H^1(C,\mathcal{M}(c)) = 0$ for all c in C. Just take $\deg \mathcal{M}$ to be maximum such that $H^1(C,\mathcal{M}) \neq 0$. Now

$$\delta_c : \Gamma(C,\mathcal{M}(c)|_c) \to H^1(C,\mathcal{M}) \to H^1(C,\mathcal{M}(c)) = 0$$

must be an isomorphism for each c because $\Gamma(C,\mathcal{M}(c))|_c)$ is one dimensional. Thus $H^1(C,\mathcal{M})$ is one dimensional and

$$\delta : \mathcal{M} \otimes \Omega_C^{\otimes -1} \to H^1(C,\mathcal{M}) \otimes \mathcal{O}_C$$

is an isomorphism. Thus $\Omega_C \approx \mathcal{M}$ and $\delta : \mathcal{O}_C \to H^1(C,\Omega_C) \otimes \mathcal{O}_C$ is an isomorphism. Hence $\delta(1) = e \otimes 1$ where e is a canonical generator of the line $H^1(C,\Omega_C)$. Thus we have proven

Lemma 8.5.1. *$H^1(C,\Omega_C)$ is canonically isomorphic to k.*

Now let $\mathcal{F}$ be an arbitrary locally free coherent sheaf on C. If $\alpha \in \mathrm{Hom}(\mathcal{F},\Omega_C)$ then we have a linear transformation,

$$H^1(C,\alpha) : H^1(C,\mathcal{F}) \to H^1(C,\Omega_C) = k.$$

Thus we have a mapping

$$H^1 : \mathrm{Hom}(\mathcal{F},\Omega_C) \to \mathrm{Hom}_k(H^1(C,\mathcal{F}), H^1(C,\Omega_C)) \cong H^1(C,\mathcal{F})\hat{}.$$

Theorem 8.5.2. (Serre duality.) *H^1 is an isomorphism between $\mathrm{Hom}(\mathcal{F},\Omega_C)$ and $H^1(C,\mathcal{F})\hat{}$.*

Proof. The proof relies on the following

Lemma 8.5.3. *Assume that we have a commutative diagram*

$$
\begin{array}{ccc}
\delta: \Omega_C \otimes_{\mathcal{O}_C} \Omega_C^{\otimes -1} & \xrightarrow{\approx} & H^1(C,\Omega_C) \otimes_k \mathcal{O}_C \\
\uparrow{\scriptstyle \alpha \otimes 1} & & \uparrow{\scriptstyle \lambda \otimes 1} \\
\delta: \mathcal{F} \otimes_{\mathcal{O}_C} \Omega_C^{\otimes -1} & \longrightarrow & H^1(C,\mathcal{F}) \otimes_k \mathcal{O}_C
\end{array}
$$

where $\alpha \in \mathrm{Hom}(\mathcal{F},\Omega_C)$ and $\lambda : H^1(C,\mathcal{F}) \to H^1(C,\Omega_C)$ is a linear transformation. Then α determines λ and conversely.

We will first see how to prove the theorem. Given $\alpha \in \mathrm{Hom}(\mathcal{F},\Omega_C)$ take $\lambda = H^1(C,\alpha)$. Then we have a commutative diagram by functionality of δ. Then the lemma implies that α is determined by λ. Hence H^1 is injective. To show that H^1 is surjective, take λ in $\mathrm{Hom}_k(H^1(C,\mathcal{F}) \to H^1(C,\Omega_C))$. As the top arrow is an isomorphism we may find α such that the diagram commutes. Now we have two commutative diagrams

(α, λ) and $(\alpha, H^1(\alpha))$. Thus the lemma implies that $\lambda = H^1(\alpha)$. So H^1 is surjective.

Proof of Lemma 8.5.3.   First we will do, "λ determines α". This is trivial as the top arrow is injective.

Next we do, "α determines λ". By the above diagram α determines $\lambda| \sum_{c \in C} \operatorname{im} \delta|_c$. We need to see that the $\operatorname{im} \delta|_c$ span $H^1(C, \mathcal{F})$. To do this it is enough to find distinct points $c_1, \ldots, c_n$ on C such that $H^1(C, \mathcal{F}(\sum c_i)) = 0$. By Lemma 8.4.2 this will follow if

$$\Gamma(C, \Omega_C \otimes \mathcal{F}^{\hat{}}(-\sum c_i)) = 0.$$

Now $\Gamma(C, \Omega_C \otimes \mathcal{F}^{\hat{}})$ is a finite dimensional vector space. If this space is not zero, let β be a non-zero section of $\Omega_C \otimes \mathcal{F}^{\hat{}}$. Take c_1 to be a point in the dense open subset where $\beta(c_1) \neq 0$. Then $\Gamma(C, \Omega_C \otimes \mathcal{F}^{\hat{}}(-c_1))$ is a smaller space. We just continue the same procedure until $\Gamma(C, \Omega_C \otimes \mathcal{F}^{\hat{}}(-\sum c_i)) = 0$.   □

Now we come to the gem.

Theorem 8.5.4. (Weil–Riemann–Roch.)   *If $\mathcal{F}$ is a locally free coherent sheaf on a smooth complete curve C, then*

$$\dim_k \Gamma(C, \mathcal{F}) - \dim_k \operatorname{Hom}(\mathcal{F}, \Omega_C) = \deg(\det \mathcal{F}) + (\operatorname{rank} \mathcal{F})(1 - g)$$

where g is the genus of C.

Proof.   Just combine Theorem 8.5.2 and Theorem 8.4.1.   □

8.6 First applications of the Riemann–Roch theorem

Let C be a smooth complete curve of genus g.

Lemma 8.6.1.

(a) $\deg \Omega_C = 2g - 2$ and
(b) $\dim \Gamma(C, \Omega_C) = g$.

Proof.   By Theorem 8.5.2, $\Gamma(C, \Omega_C) = \operatorname{Hom}(\mathcal{O}_C, \Omega_C)$ is dual to the g dimensional space $H^1(C, \mathcal{O}_C)$. Thus (b) is true. As $H^1(C, \Omega_C) = k, \chi(\Omega_C) = g - 1 = \deg(\Omega_C) + 1 - g$ by Theorem 8.4.3. Thus (a) follows by arithmetic.   □

Let $f : C \to D$ be a separable morphism where D is another smooth complete curve D.

Proposition 8.6.2. (Riemann-Hurwitz)

$$2 \operatorname{genus}(C) - 2 = (\deg f)(2 \operatorname{genus}(D) - 2) + \deg(\operatorname{div} \Omega_{C/D}).$$

Proof. By Lemma 7.6.2(a) we have an isomorphism $\Omega_C = f^*\Omega_D(\operatorname{div} \Omega_{C/D})$. Thus $\deg \Omega_C = (\deg f)\deg \Omega_D + \deg(\operatorname{div} \Omega_{C/D})$. Thus Proposition 8.6.2 follows from Lemma 8.6.1. $\quad\square$

Assume now that char $k \neq 2$ and $\deg f = 2$. Then f is a double cover.

Corollary 8.6.3. *In this case* $\operatorname{genus}(C) = \operatorname{genus}(D) + \frac{1}{2}$ *(number of ramification points).*

Proof. See Lemma 7.6.2 to compute $\operatorname{div} \Omega_{C/D}$. $\quad\square$

Exercise 8.6.4. Construct all double covers of $\mathbf{P}^1$ at least if char $k \neq 2$. Compute their genus and show that any non-negative integer is the genus of some curves.

These simple double covers of $\mathbf{P}^1$ are the infamous *hyperelliptic curves*.

Exercise 8.6.5. Show that the curve C is hyperelliptic iff there is an invertible sheaf $\mathcal{L}$ on C such that $\deg \mathcal{L} = 2$ and $\dim \Gamma(C, \mathcal{L}) \geq 2$. We have inequality iff $C = \mathbf{P}^1$. Any curve of genus 2 is hyperelliptic.

In the purely inseparable case we have

Lemma 8.6.6. *Assume that* $f : C \to D$ *is purely inseparable; then* $\operatorname{genus}(C) = \operatorname{genus}(D)$.

Proof. $\operatorname{genus}(C) = \dim_k H^1(C, \mathcal{O}_C) = \dim_k H^1(D, f_*\mathcal{O}_D) = \dim_k H^1(D, \mathcal{O}_D^{1/q}) = \dim_k H^1(D, \mathcal{O}_D) = \operatorname{genus}(D)$ by Lemma 7.6.3. $\quad\square$

Let $f : C \to D$ be general.

Proposition 8.6.7. *If* $\operatorname{genus}(D) \geq 1$ *then* $\operatorname{genus}(C) \geq \operatorname{genus}(D)$ *with equality only if* $\operatorname{genus}(D) = 1$.

Proof. Factor f as $C \xrightarrow{g} E \xrightarrow{h} D$ where g is purely inseparable and h is separable. By Lemma 8.6.6 we may assume that $f = h$; i.e. f is separable. Then we have by Proposition 8.6.2

$$2 \operatorname{genus}(C) - 2 \geq (2 \operatorname{genus}(D) - 2)(\deg f)$$

as div $\Omega_{C/D}$ is effective. As deg $f \geq 1$ the result follows. $\square$

Corollary 8.6.8. (Lüroth)   *There is no non-constant morphism* $\mathbb{P}^1 \to$
D where genus(D) > 0.

Proof. Genus($\mathbb{P}^1$) $= 0$. $\square$

8.7 Residues and the trace homomorphism

Let ω be a rational differential on a smooth complete curve C. Let c be
a point of C. We want to define the *residue* $\mathrm{Res}_c(\omega)$ of ω at c. Consider
the isomorphism $r : H^1(C, \Omega_C) \to k$. $\mathrm{Res}_c(\omega) = r$ (cohomology class of
the principal part of ω at c).

We want to show that $\mathrm{Res}_c(\omega)$ is the Cauchy residue of ω at c by
checking

Theorem 8.7.1.

(a) Res_c: Rat $\Omega_C \to k$ *is* k-*linear.*

(b) $\mathrm{Res}_c(\omega) = 0$ *if* ω *is regular at* c.

(c)   *If* π *is a parameter at* c, $\mathrm{Res}_c(\frac{1}{\pi} d\pi) = 1$ *and* $\mathrm{Res}_c(\frac{1}{\pi^n} d\pi) = 0$ *if*
$n > 1$.

Proof. (a) is evident and (b) follows because the principal part of ω at
c is zero. It suffices to prove (c). We have to compute the isomorphism
$\pi_{2*}(\pi_1^* \Omega_C(\Delta)|_\Delta) \cong \Omega_C \otimes \Omega_C^{\otimes -1} \widetilde{\cong} \mathcal{O}_C$. Locally around c, $d\pi$ is a basis
on Ω_C. An expression in $\pi_{2*}(\pi_1^* \Omega_C(\Delta)|_\Delta)$ near (c, c) has the form $d\tau \times$
$(\pi(c_1) - \pi(c_2))^{-1}$ times a regular function $f(c_1, c_2)$ modulo functions
vanishing on the diagonal. This corresponds to $(d\tau \otimes \frac{\partial}{\partial \tau}) \cdot f(c_1, c_1)$ in
$\Omega_C \otimes \Omega_C^{\otimes -1}$ which goes to $f(c_1, c_1)$ in $\mathcal{O}_C$. Thus $\delta(d\pi(\pi(c_1) - \pi(c_2))^{-1}) =$
$e \otimes 1$. Evaluation at c then δ_c (cohomology class of $\frac{d\pi}{\pi}$ at c) goes to e.
The first point of (c) follows because $r(e) = 1$.

The rest of (c) follows by continuity as in Cauchy's original theo-
rem. Consider the sequence of sheaves $0 \to \pi_1^* \Omega_C (\to \pi_1^* \Omega_C(\Delta_{1,2} + \ldots +$
$\Delta_{1,n+1}) \to \pi_1^* \Omega_C(\Delta_{1,2} + \ldots + \Delta_{1,n+1})|_{(\Delta_{12} + \ldots + \Delta_{1,n+1})} \to 0$ on C^{n+1}
where $\Delta_{i,j}$ is the diagonal $\{(c_1, \ldots, c_{n+1})|c_i = c_j\}$.
Thus we have a connecting homomorphism

$$\delta : \pi_{2 \ldots n+1 *}(\pi_1^* \Omega_C(\Delta_{1,2} + \ldots + \Delta_{1,n+1})|_{(\Delta_{12} + \ldots + \Delta_{1,n+1})}) \to$$

$$H^1(C^n, \pi_{2 \ldots n+1 *}\pi_1^* \Omega_C) = H^1(C, \Omega_C) \otimes_k \mathcal{O}_{C^n}.$$

Let $d = (c_2, \ldots, c_{n+1})$ be a point of C^n. Then the value of δ at d

is the boundary $\delta_{\sum c_i} : \Omega_C(\sum c_i)|_{\sum c_i} \to H^1(C, \Omega_C)$ of the sequence $0 \to \Omega_C \to \Omega_C(\sum c_i)| \to \Omega_C(\sum c_i)|_{\sum c_i} \to 0$.

Thus limit $\delta_{\sum c_i} = \delta_{nc}$ and if the c_i's are distinct $\delta_{\sum c_i} = \sum_i \delta_{c_i}$.

Consider δ of $\omega = d\pi / \prod_{2 < i \leq n+1} (\pi(x_1) - \pi(x_i))$. Which is a basic section of $\pi_1^* \Omega_C(\Delta_{1,2} + \ldots + \Delta_{1,n+1})$. We assume that $\pi(C_2), \ldots, \pi(C_n + V)$ are distinct. Then $\delta(\omega)$ at $(c, \ldots, c) =$ cohomology classes of $\frac{d\pi}{\pi^n}$ at c and if the $\pi(c_r)$ are distinct, $\delta(\omega)$ at $(c_1, \ldots, c_{n+1}) = \sum_i$ cohomology classes of $\frac{d\pi}{\prod \pi - \pi(c_i)}$ at c_i. Thus

$$\operatorname{Res}_c \frac{d\pi}{\pi^n}(\text{class of } \frac{d\pi}{\pi^n}) = \left(\varinjlim \sum_i \text{cohomology class of } \frac{d\pi}{\prod(\pi - \pi(c_i))} \right)$$

$$= \varinjlim \sum_i \operatorname{Res}_{c_i} \left(\frac{d\pi}{\prod(\pi - \pi(c_i))} \right)$$

$$= \varinjlim \sum_i \prod_{j \neq i} \frac{1}{(\pi(c_i) - \pi(c_j))}.$$

But the Lagrange identity gives

$$1 = \sum_i \frac{(\prod_{i \neq j} X - \pi(c_j))}{\prod_{i \neq j} \pi(c_i) - \pi(c_j)} \quad \text{where } X \text{ is an indeterminate.}$$

Taking the coefficient of X^{n-1} we have

$$0 = \sum_i \prod_{j \neq i} \frac{1}{\pi(c_i) - \pi(c_j)}.$$

Thus $\operatorname{Res}_c \frac{d\pi}{\pi^n} = 0.$ $\qquad \square$

Exercise 8.7.2. Let $p_1, \ldots, p_d$ be principal parts of rational differentials at points $c_1, \ldots, c_d$. Then $\sum_i p_i$ is the principal part of a rational differential iff $\sum_i \operatorname{Res}_{c_i} p_i = 0$.

Now we come to the trace mapping. Let $f : C \to D$ be a non-constant morphism between smooth complete curves. As f is affine,

$$H^1(C, \Omega_C) \cong H^1(D, f_* \Omega_C).$$

Thus our canonical isomorphism $H^1(C, \Omega_C) \xrightarrow{\approx} k$ gives a linear functional $\lambda : H^1(D, f_* \Omega_C) \to k$. By Theorem 8.5.2 we have a corresponding $\mathcal{O}_D$-homomorphism

$$\operatorname{Tr} : f_* \Omega_C \to \Omega_D.$$

We want to compute Tr locally when f is separable.

If f is separable we have an $\mathcal{O}_D$-homomorphism $\operatorname{Tr}' : \mathbf{Rat}(f_* \Omega_C) \to$

Rat Ω_D defined by $\mathrm{Tr}'(g \cdot f^{-1}\omega) = \mathrm{Tr}(g) \cdot \omega$ where g is a rational function on C and ω is a non-zero rational differential on D. One easily checks that Tr' is well-defined by the formula.

Theorem 8.7.3. $\mathrm{Rat}(\mathrm{Tr}) = \mathrm{Tr}'$ *if f is separable.*

Proof. Tr is characterized by the property that for all points d in an open subset U of D, $\mathrm{Tr}: f_*\Omega_C(d)|_d \to \Omega_D(d)|_d$ maps the principal part of $\frac{d\pi}{\pi}$ at c to the principal part of $\frac{d\sigma}{\sigma}$ at d where $f(c) = d$ and π is a parameter at c and σ is a parameter at d.

Let U be the open subset over which f is unramified. Then Tr' takes $f_*\Omega_C|_U$ into $\Omega_D|_U$ because if σ is a parameter at a point u of U, $\mathrm{Tr}'(gf^*d\sigma) = \mathrm{Tr}(g)d\sigma$ which is regular at u if g is regular on $f^{-1}u$. To check that $\mathrm{Tr}' = \mathrm{Tr}|_u$ we need to compute when f is regular on $f^{-1}U$. $\mathrm{Tr}'(g\frac{f^*d\sigma}{f^*\sigma})|_u = \mathrm{Tr}(g)|_u \frac{d\sigma}{\sigma}|_u = (\sum_{x \in f^{-1}u} g(x)) \frac{d\sigma}{\sigma}|_u$. This shows what we wanted. $\qquad\square$

Exercise 8.7.4. Let $\mathcal{F}$ be a local free coherent sheaf on a smooth complete curve C.

If P is a principal part for $\mathcal{F}$, then P is the principal part of a rational section of $\mathcal{F}$ if and only if

$$\mathrm{Res}(< P, x >) = 0 \text{ for all } x \text{ in } \Gamma(C, \mathcal{F}^\vee \otimes_{\mathcal{O}_C} \Omega_C).$$

(This describes the obstruction to solving the Cousin problem of $\mathcal{F}$.)

9

General cohomology

Let n be a positive integer. We want to consider the cohomology $H^i(\mathbf{A}^n - \{0\}, \mathcal{O}_{\mathbf{A}^n})$ of the punctured affine space $\mathbf{A}^n$. Fortunately this cohomology is a $k[\mathbf{A}^n] = k[X_1, \ldots, X_n]$-module under multiplication. So we want to write an isomorphism of it with a concrete $k[\mathbf{A}^n]$-module.

Proposition 9.1.1.

(a) $H^i(\mathbf{A}^n - \{0\}, \mathcal{O}_{\mathbf{A}^n}) = 0$ *unless* $i = 0$ *or* $n - 1$,

(b) *if* $n = 1$, $H^0(\mathbf{A}^n - \{0\}, \mathcal{O}_{\mathbf{A}^n}) = k[X_1, X_1^{-1}]$,

(c) *if* $n > 1$, $H^0(\mathbf{A}^n - \{0\}, \mathcal{O}_{\mathbf{A}^n}) = k[X_1, \ldots, X_n]$ *and* $H^{n-1}(\mathbf{A}^n - \{0\}, \mathcal{O}_{\mathbf{A}^n}) = \bigoplus_{\substack{p \in \mathbb{Z}^n \\ p_i \leq -1}} k X_1^{p_1} \ldots X_n^{p_n}$ *where the module structure is the obvious one.*

Proof. If $n = 1$, $\mathbf{A}^1 - \{0\} = D(X_1)$ is affine. Hence its higher cohomology groups are zero and its section is $k[\mathbf{A}^1]_{(X_1)}$. We will proceed by induction on n. If $n > 1$ we have the exact sequence

$$0 \to \mathcal{O}_{\mathbf{A}^n - \{0\}} \to_{D(X_n)} \mathcal{O}_{\mathbf{A}^n - \{0\}} \to \bigoplus_{p_n \leq -1} \mathcal{O}_{\mathbf{A}^{n-1} - (0)} X_n^{p_n} \to 0.$$

As $D(X_n)$ is affine and the inclusion of $D(X_n)$ in $\mathbf{A}^n - \{0\}$ is affine the middle sheaf has no higher cohomology and its section is $k[X_1,\ldots,$ $X_n, X_n^{-1}]$. Looking at the long exact sequence, if $n = 2$, we have

$$0 \to H^0(\mathbf{A}^2 - \{0\}, \mathcal{O}_{\mathbf{A}^2}) \to k[X_1,\ldots,X_2,X_2^{-1}] \xrightarrow{\epsilon}$$

$$\sum_{p_2 \leq -1} k[X_1, X_1^{-1}] X_2^{p_2} \to H^1(\mathbf{A}^2 - \{0\}, \mathcal{O}_{\mathbf{A}^2}) \to 0$$

is exact and $H^i(\mathbf{A}^2 - \{0\}, \mathcal{O}_{\mathbf{A}^2}) = 0$ if $i \geq 2$. The mapping ϵ erases non-negative powers of X_2. Thus the result is clear in this case. If $n > 2$ we have an exact sequence

$$0 \to H^0(\mathbf{A}^n - \{0\}, \mathcal{O}_{\mathbf{A}^n}) \to k[X_1,\ldots,X_n,X_n^{-1}] \xrightarrow{\epsilon}$$

$$\sum_{p_n \leq -1} k[X_1,\ldots,X_{n-1}] X_n^{p_n} \to H^1(\mathbf{A}^n - \{0\}, \mathcal{O}_{\mathbf{A}^n}) \to 0.$$

As ϵ is surjective $H^1(\mathbf{A}^n - \{0\}, \mathcal{O}_{\mathbf{A}^n}) = 0$. Furthermore we have the boundary if $i > 1$

$$H^{i+1}(\mathbf{A}^n - \{0\}, \mathcal{O}_{\mathbf{A}^n}) \overset{\delta}{\underset{\approx}{}} \sum_{p_n \leq -1} H^i(\mathbf{A}^{n-1} - \{0\}, \mathcal{O}_{\mathbf{A}^{n-1}}) X_n^{p_n}.$$

Thus the result follows by induction. □

Consider the projection $\pi : \mathbf{A}^{n+1} - \{0\} \to \mathbf{P}^n$. This is affine. Furthermore, $\pi_* \mathcal{O}_{\mathbf{A}^{n+1}-\{0\}} = \bigoplus_{d \in \mathbf{Z}} \mathcal{O}_{\mathbf{P}^n}(d)$. Thus $H^i(\mathbf{P}^n, \mathcal{O}_{\mathbf{P}^n}(d))$ is the term of degree d in $H^i(\mathbf{A}^{n+1} - \{0\}, \mathcal{O}_{\mathbf{A}^{n+1}})$. We get

Corollary 9.1.2.

(a)   *The cohomology* $H^i(\mathbf{P}^n, \mathcal{O}_{\mathbf{P}^n}(d)) = 0$ *unless $i = 0$ or n.*

(b) $H^0(\mathbf{P}^n, \mathcal{O}_{\mathbf{P}^n}(d)) = k[X_0,\ldots,X_n]_{\text{deg. } d}$

(c) $H^n(\mathbf{P}^n, \mathcal{O}_{\mathbf{P}^n}(d)) = k[X_0^{-1},\ldots,X_n^{-1}] X_0^{-1} \ldots X_n^{-1}_{\text{deg. } d}$

(d) $H^n(\mathbf{P}^n, \mathcal{O}_{\mathbf{P}^n}(-n-1))$ *is one dimensional and the multiplication* $H^0(\mathbf{P}^n, \mathcal{O}_{\mathbf{P}^n}(d)) \times H^n(\mathbf{P}^n, \mathcal{O}_{\mathbf{P}^n}(-n-1-d)) \to H^n(\mathbf{P}^n, \mathcal{O}_{\mathbf{P}^n}(-n-1))$ *is non-degenerate.*

Proof.   The only new fact is (d). Here $H^n(\mathbf{P}^n, \mathcal{O}_{\mathbf{P}^n}(-n-1)) = kX_0^{-1} \ldots X_n^{-1}$ and the statement about the multiplication is obvious. □

9.2 Čech cohomology and the Künneth formula

Let X be a topological space with open subsets, $U_1,\ldots,U_d$. If $\mathcal{F}$ is an abelian sheaf we will define a complex

$$\check{C}^*(\mathcal{F}) = \check{C}^0(\mathcal{F}) \xrightarrow{\delta^0} \check{C}^*(\mathcal{F}) \to \ldots$$

with an augmentation $\Gamma(X, \mathcal{F}) \to \check{C}^*(\mathcal{F})$. By definition

$$\check{C}^n(\mathcal{F}) = \bigoplus_{i_0 < \ldots < i_n} \Gamma(U_{i_0} \cap \ldots \cap U_{i_n}, \mathcal{F})$$

and the differential $\hat{\delta}(\alpha) = (\beta_{i_0 < \ldots < i_{n+1}})$ where

$$\beta_{i_0 < \ldots < i_{n+1}} = \sum_{0 \leq j \leq n} (-1)^j \alpha_{i_0 < \ldots < i_j < \ldots < i_n} |_{U_{i_0} \cap \ldots \cap U_{i_{n+1}}}.$$

As usual $\delta^2 = 0$ and $\check{C}^*(X, \mathcal{F})$ is a complex.

The *Čech cohomology* $\check{H}^i(X, \mathcal{F})$ is the i-th homology of the complex $\check{C}^*(X, \mathcal{F})$.

Lemma 9.2.1. *If $U_P = X$, then the complex*

$$0 \to \Gamma(X, \mathcal{F}) \overset{\epsilon}{\to} \check{C}^*(\mathcal{F})$$

is homotopically trivial and hence exact.

Proof. We need to define an operator $\Gamma(X, \mathcal{F}) \overset{k_0}{\leftarrow} \check{C}^0(\mathcal{F}) \overset{k_1}{\leftarrow} \check{C}^1(\mathcal{F}) \leftarrow \ldots$ such that

$$(*) \delta k + k \delta = \text{identity}$$

where $\delta_{-1} = \epsilon$, $k_0((\alpha_i)) = \alpha_p$ and

$$k_n(\alpha_{i_0 < p < i_{n-1}}) = (\alpha_{i_0 < \ldots < j_{n-1}})$$

and k_n is zero otherwise. One checks that $(*)$ is true. $\qquad\square$

Proposition 9.2.2. *Let $\mathcal{F}$ be a quasi-coherent sheaf on a separated variety X. Let $U_0, \ldots, U_n$ be open affine subsets of X which cover X. Then $H^i(X, \mathcal{F}) \overset{\cong}{\to} \check{H}^i(X, \mathcal{F})$ for all i.*

Proof. We define a sheaf version of $\check{C}^*(X, \mathcal{F})$. For any open subset V of X, let $\check{C}^*(\mathcal{F})(V) = \check{C}^*(V, \mathcal{F})$ with respect to the covering $V \cap U_0, \ldots, V \cap U_n$. If we give obvious restrictions $C^*(\mathcal{F})$ is a complex of sheaves and we have the augmentation $\mathcal{F} \to \check{C}^* \mathcal{F}$. By the resolution principle it suffices to show that $\mathcal{F} \to \check{C}^* \mathcal{F}$ is a resolution and, for each i, $H^j(X, \check{C}^* \mathcal{F}) = 0$ for $j > 0$.

The first statement is local. So we may assume that $X = $ some U_p. In this case the exactness follows from Lemma 9.2.1. To show the vanishing note that $\check{C}^i(\mathcal{F}) = \bigoplus_{k_1 < \ldots < k_i} U_{k_1} \cap \ldots \cap U_{k_i} \mathcal{F}$. As X is separated, $U_{k_1} \cap \ldots \cap U_{k_i}$ is affine and its inclusion in X is affine. Thus the vanishing follows as usual. $\qquad\square$

Corollary 9.2.3.   *In the above situation*

$$H^j(X, \mathcal{F}) = 0 \ if \ j > n.$$

We also get the *Künneth formula*.

Proposition 9.2.4.   *Let $\mathcal{F}_1$ and $\mathcal{F}_2$ be quasi-coherent sheaves on separated varieties X_1 and X_2.*
Then $H^i(X_1 \times X_2, \pi_1^* \mathcal{F}_1 \otimes \pi_2^* \mathcal{F}_2) = \bigoplus_{i_1+i_2=i} H^{i_1}(X_1, \mathcal{F}_1) \otimes_k H^{i_2}(X_2, \mathcal{F}_2).$

Proof. Let $U_{1,0}, \ldots, U_{1,p}$ and $U_{2,0}, \ldots, U_{2,q}$ be open affine coverings of X_1 and X_2. Then we have a resolution $\pi_1^* \mathcal{F}_1 \otimes \pi_2^* \mathcal{F}_2 \to \pi_1^* \check{C}^*(\mathcal{F}_1) \otimes \pi_2^* \check{C}^*(\mathcal{F}_2)$. If we apply the resolution principle here as before we get $H^*(X_1, \times X_2, \pi_1^* \mathcal{F} \otimes \pi_2^*) \approx i$-homology of $\check{C}^*(X_1, \mathcal{F}_1) \otimes_k \check{C}^*(X_2, \mathcal{F}_2)$. Now the result follows from linear algebra.   □

9.3 Cohomology of projective varieties

We want to prove

Theorem 9.3.1.   *Let $\mathcal{F}$ be a coherent sheaf on a projective variety X. Then*

(a)   *the cohomology groups $H^i(X, \mathcal{F})$ are finite dimensional k-vector-spaces;*

(b)   *There exists n_0 such that $H^i(X, \mathcal{F}(n)) = 0$ if $i > 0$ and $n \geq n_0$.*

Proof. We have a closed embedding $X \subset \mathbf{P}^n$. As closed embeddings do not change cohomology we may assume that $X = \mathbf{P}^n$. We will prove the theorem by descending induction on i. If $i > n$ there is no problem because the cohomology is zero by Corollary 9.2.3. We may find an exact sequence

$$0 \to \mathcal{G} \to \bigoplus_{\text{finite}} \mathcal{O}_{\mathbf{P}^n}(p) \to \mathcal{F} \to 0 \ \text{where} \ p \ll 0.$$

Now $\mathcal{G}$ is coherent. Thus by Corollary 9.1.2 we have an isomorphism $H^i(\mathbf{P}^n, \mathcal{F}) \xrightarrow{\approx} H^i(\mathbf{P}^n, \mathcal{G})$ if $i < n$ and a surjection $\bigoplus H^n(\mathbf{P}^n, \mathcal{O}_P(p)) \to H^n(\mathbf{P}^n, \mathcal{F}) \to 0$. Thus if $i = n$ we have finite dimensionality because we know it for $\mathcal{O}_P(p)$ by Corollary 9.1.2. If $i < n$ the result follows by induction from the case of $\mathcal{G}$. This proves (a). For (b) do a descending induction on i. As the theorem is true for $\mathcal{O}_{\mathbf{P}^n}(m)$ by Corollary 9.1.2 we can just use the long exact sequences tensored by $\mathcal{O}_{\mathbf{P}^n}(m)$.   □

Let $\mathcal{F}$ be a coherent sheaf on a projective variety X. The *Euler characteristic* of $\mathcal{F}$ is the integer $\chi(\mathcal{F}) = \sum_i (-1)^i \dim_k H^i(X, \mathcal{F})$. If X is smooth, Hirzebruch gave a formula to compute $\chi(\mathcal{F})$ in terms of the geometry of X and $\mathcal{F}$. This is called the *Hirzebruch–Riemann–Roch theorem*. This result generalizes Theorem 8.5.4. Later it was generalized by Grothendieck to a theorem about higher direct images [EGA]. This interesting material is beyond the scope of this book.

Exercise 9.3.2. Show that

$$\chi(\mathcal{O}_{\mathbf{P}^n}(r)) = \frac{(n+r)!}{n! \, r!}.$$

This argument was so easy we can generalize it and prove

Theorem 9.3.3. *Let X be a projective variety and Y be a variety. If $\mathcal{F}$ is a coherent sheaf on $X \times Y$ then*

(a)   *the higher direct images $R^i \pi_{Y*} \mathcal{F}$ are coherent sheaves on Y,*

(b)   *There exists n_0 such that $R^i \pi_{Y*} \mathcal{F}(n) = 0$ for $i > 0$ and $n \geq n_0$.*

Proof. The result is local on Y. So we may assume that Y is affine. Also let $X = \mathbf{P}^n$. In this case we will later remark that we have an exact sequence

$$(*) \qquad 0 \to \mathcal{G} \to \bigoplus_{\text{finite}} (\pi^*_{\mathbf{P}^n}(p)) \to \mathcal{F} \to 0 \text{ as before.}$$

Then we need to know that $R^i \pi_{Y*}(\pi^*_{\mathbf{P}^n}(p)) = H^i(\mathbf{P}^n, \mathcal{O}_{\mathbf{P}^n}(p)) \otimes_k \mathcal{O}_Y$ (Lemma 8.3.3) and $R^i \pi_{Y*}(\mathcal{F}) = (H^i(X \times Y, \mathcal{F}))^\sim = 0$ if $i > n$ (Corollary 9.2.3). The same argument applies to replace cohomology by direct images. □

Thus we need to better understand coherent sheaves on $\mathbf{P}^n \times Y$ where Y is affine. The whole theory of Section 5.4 generalizes. Consider the graded ring $k[\mathbf{A}^{n+1}] \otimes_k k[Y] = B$. Then one constructs a quasi-coherent sheaf $\tilde{M}$ on $\mathbf{P}^n \times Y$ for any graded B-module such that all quasi-coherent sheaves are constructed this way and any coherent sheaves come from a finitely generated graded B-module. Everything works as before. Let $\pi : \mathbf{A}^{n+1} - \{0\} \times Y \to \mathbf{P}^n \times Y$ be the projection. Then $\tilde{M}$ is the degree zero part of $\pi_*(\tilde{\tilde{M}}|_{\mathbf{A}^{n+1} - \{0\} \times Y})$ where $\tilde{\tilde{M}}$ is the quasi-coherent sheaf on $\mathbf{A}^{n+1} \times Y = \mathrm{Spec}(B)$ associated to M. There are no new ideas in this generalization.

9.4 The direct images of flat sheaves

Let $f : X \to Y$ be a morphism. Let $\mathcal{F}$ be a quasi-coherent sheaf on X and $\mathcal{G}$ be a quasi-coherent sheaf on Y. We have the sheaves

$$R^i f_*(\mathcal{F} \otimes f^*\mathcal{G}) \text{ on } Y.$$

In this section we will study how these sheaves depend on $\mathcal{G}$ in a special situation.

The morphism $f : X \to Y$ will be *assumed* to have the form $\pi_Y :$ $Z \times Y \to Y$ where Z is a projective variety. The sheaf $\mathcal{F}$ will be *assumed* to be flat over Y (to be defined) and coherent. In this section we will prove

Theorem 9.4.1. (Grothendieck). *Any point of Y is contained in an open neighborhood U such that we have a complex $0 \to K^0 \to K^1 \to \ldots \to K^n \to 0$ where each K^i is a free $\mathcal{O}_U$-module of finite rank and a natural isomorphism $R^i f_*(\mathcal{F} \otimes f^*\mathcal{G}|_U) \approx i$-homology sheaf of the complex $K^* \otimes_{\mathcal{O}_U} \mathcal{G}|_U$.*

In the next section we will explain some of the meaning of this theorem.

We begin with the definition of $\mathcal{F}$ being *flat* over Y. It means that for all points x of X the stalk $\mathcal{F}_x$ is a flat $\mathcal{O}_{Y,f(x)}$-module via $f^* :$ $\mathcal{O}_{Y,f(x)} \to \mathcal{O}_{X,x}$ where a module M over a ring A is flat if $M \otimes_A G$ is an exact functor of A-modules G (and not just right exact).

Lemma 9.4.2. *Let U and V be open affine subsets of X and Y such that $f(U) \subset V$. If $\mathcal{F}$ is flat over Y, then $\Gamma(U, \mathcal{F})$ is a flat $k[V]$-module.*

Proof. Let G be $k[V]$-module. Then $\Gamma(U, \mathcal{F}) \otimes_{k[V]} G = \Gamma(U, \mathcal{F} \otimes f^*\tilde{G})$. Thus this functor is the composition of the three exact functors $\simeq, \mathcal{F} \otimes$ f^*- and $\Gamma(U,-)$. Therefore it is exact. $\square$

Next we begin the

Proof of Theorem 9.4.1. The result is local on Y. So we may assume that Y is affine. By the equivalence of cohomology and direct images we need to compute $H^i(X, \mathcal{F} \otimes f^*\tilde{G})$ where G is $k[Y]$-module. As X is separated we may use Čech cohomology. Let $X = U_0 \cup \ldots \cup U_n$ be an open affine cover of X. Then by Proposition 9.2.2 $\check{H}^i(X, \mathcal{F} \otimes f^*\tilde{G}) = H^i(X, \mathcal{F} \otimes f^*\tilde{G})$. Let $\check{C}^*(\mathcal{F}) \otimes f^*\tilde{G}$ be the Čech complex. Then $\check{C}^*(\mathcal{F}) \otimes f^*\tilde{G} = C^*(\mathcal{F}) \otimes_{k[Y]} G$ by the obvious calculation $\Gamma(U_{i_1} \cap \ldots \cap$

$U_{i_j}, \mathcal{F} \otimes f^* \tilde{G}) = \Gamma(U_{i_1} \cap \ldots \cap U_{i_j}, \mathcal{F}) \otimes_{k[Y]} G$. Thus $H^i(X, \mathcal{F} \otimes f^* \tilde{G})$ is the i-homology group of the complex $\check{C}^*(\mathcal{F}) \otimes_{k[Y]} G$.

We would be done if $\check{C}^*(\mathcal{F})$ were a complex of free $k[Y]$-modules of finite rank but this is not the case. We know that $\check{C}^i(\mathcal{F})$ is $k[Y]$-flat and the cohomology of $\check{C}^*(\mathcal{F})$ consists of $k[Y]$-modules of finite type by Theorem 9.3.3(a).

Lemma 9.4.3. *There exist a complex $0 \to L^0 \to L^1 \to \ldots \to L^n \to 0$ of $k[Y]$-modules of finite type where $L^1, \ldots, L^n$ are free and a homomorphism $\psi : L^* \to \check{C}^*(\mathcal{F})$ which induces an isomorphism on cohomology.*

Proof. We want to construct the L^i by descending induction so that we have a map of complexes

$$
\begin{array}{ccccccccc}
0 & \longrightarrow & L^i & \longrightarrow & L^{i+1} & \longrightarrow & \ldots L^n & \longrightarrow & 0 \\
 & & \downarrow & & \downarrow & & \downarrow & & \downarrow \\
 & \check{C}^{i-1}(\mathcal{F}) & \to & \check{C}^i(\mathcal{F}) & \to & \check{C}^{i-1}(\mathcal{F}) & \ldots & \check{C}^n(\mathcal{F}) & \to & 0
\end{array}
$$

inducing an isomorphism on cohomology of level $> i$ and a surjection on level i. Thus if $i - 1 > 0$ we can easily find L^{i-1} because $H^i(C, \mathcal{F})$ is a finite type $k[Y]$-module and $k[Y]$ is noetherian. If $i = 0$ take $L^0 = \mathrm{Ker}(\mathrm{Ker}\ L^1 \to H^1(X, \mathcal{F})) \oplus H^0(X, \mathcal{F})$. This maps to L^1 by the first projection and to $\check{C}^0(\mathcal{F})$ by the second. Thus we have solved the problem. $\square$

To prove the theorem we want to take $K^* = \tilde{L}^*$. To proceed we need to prove

(a) K^0 is locally free (then we can localize further to make K^0 free) and

(b) $\psi \otimes_{k[Y]} G$ induces an isomorphism on homology for all modules G.

The idea is to consider the mapping cone of ψ. This is a complex $C^*(f)$ together with an exact sequence $0 \to \check{L}^{*-1} \to C^*(f) \to \check{L} \to 0$ of complexes. Thus we have a long sequence of homologies

$(*)$ $\to H^{i-1}(\check{L}^*_{(F)}) \to H^i(C^*(f)) \to H^i(L^*) \overset{\pm\psi}{\to} H^i(\check{C}^*(\mathcal{F})) \to \ldots$.

By definition

$$ C^i(f) = \check{C}^{i-1}(\mathcal{F}) \oplus L^i $$

and

$$ \delta^i(\alpha, \beta) = (\delta^{i-1}\alpha + (-1)^i \psi(\beta), \delta^i(\beta)). $$

One of these gives a complex and the obvious mapping gives the exact sequence and ∗ holds by computing the snake homomorphism explicitly.

On (∗) by construction the $\pm\psi$ are all isomorphisms so $C^*(f)$ is exact. Now we have a finite exact sequence

$$0 \to L^0 = C^0 \to C^0(\mathcal{F}) \oplus L^1 \to$$

where everything except L^0 is $k[Y]$-flat. L^0 is a flat $k[Y]$-module.

Therefore $C^*(f) \otimes_{k[Y]} G$ is exact for any $k[Y]$-module G. Therefore by $\star \otimes G$, $H^i(L^* \otimes_{k[Y]} G) \to H^i(\tilde{C}^*(\mathcal{F}) \otimes_{k[Y]} G)$ is an isomorphism for all i. This proves (b).

For (a) we just need to apply

Lemma 9.4.4. *If L is a flat $k[Y]$-module of finite type then $\tilde{L}$ is locally free.*

Proof. Let y be point of Y. Let $\ell_1, \ldots, \ell_n$ be elements of L which give a basis of $\tilde{L}|_y$. Then $\ell_1, \ldots, \ell_n$ span the coherent sheaf $\tilde{L}$ in a neighborhood V of y. Thus we have an exact sequence $0 \to \mathcal{G} \to \bigoplus \mathcal{O}_U \to \tilde{L}|_U \to 0$ where $\mathcal{G}$ is coherent. As $\tilde{L}|_U$ is flat, $\text{Tor}^1(L, \mathcal{O}_U) = 0 \to \mathcal{G}|_y \to \bigoplus \mathcal{O}_U|_y \overset{+\psi}{\to} \tilde{L}|_u \to 0$ is exact. By construction ψ is an isomorphism. Thus $\mathcal{G}|_y$ so $\mathcal{G}$ is zero in a neighborhood of y. Hence $\tilde{L}$ is locally free.   □

9.5 Families of cohomology groups

We continue with the assumption of the last section.

Let y be a point of Y. Let $X_y = f^{-1}(y)$ and $\mathcal{F}_y = \mathcal{F}|_{X_y}$. We want to study how the cohomology of $\mathcal{F}_y$ changes as we vary y.

Proposition 9.5.1.

(a)   The Euler characteristic $\chi(\mathcal{F}_y)$ is a locally constant function of y.
(b)   For all i $\dim_k H^i(X, \mathcal{F}_y)$ is an upper-semicontinuous function of y.

Proof.   The key remark is that $\mathcal{F}|_y = \mathcal{F} \otimes f^* \mathcal{O}_{\{y\}}$. This makes the connection with the material of the last section. This proposition is local on y. So by Theorem 9.4.1 we may assume that we have the complex

$$0 \to K^0 \overset{\alpha^0}{\to} K^1 \to \ldots \to K_n \to 0$$

of free $\mathcal{O}_y$-modules of finite rank such that the i-homology of $K^*|_y$ is $\cong H^i(X_y, \mathcal{F}_y)$ for all y in Y. Here we identify $H^i(X_y, \mathcal{F}_y)$ with the skyscraper sheaf $R^i f_*(\mathcal{F}_y)$. By easy linear algebra $\chi(\mathcal{F}_y) = \sum(-1)$ dim $H^i(X_y, \mathcal{F}_y) = \sum(-1)$ dim $H^i(K^*|_y) = \sum(-1)^i$ dim $K^*|_y = \sum(-1)^i$ rank K^i. Thus (a) is true because $\chi(\mathcal{F}_y)$ is constant.

For (b) we have the exact sequence

$$0 \to \operatorname{Im}(\alpha^{i-1}|_y) \to \operatorname{Ker}(\alpha^i|_y) \to H^i(K^*|_y) \to 0$$

and $H^i(K^*|_y) = H^i(X_y, \mathcal{F}_y)$.

Thus dim $H^i(X_y, \mathcal{F}_y)$ = dim $K^i|_y$−rank$(\alpha^i|_y)$−rank$(\alpha^{i-1}|_y)$ where dim $K^i|_y$ = rank K^i is constant. Thus (b) follows because the rank of a matrix of regular functions is lower-semicontinuous. □

One may ask to study this machinery in greater detail. The general theme is to look at the natural mapping

$$\psi_i(y) : (R^i f_* \mathcal{F})|_y \to H^i(X_y, \mathcal{F}_y).$$

In general $\psi_i(y)$ is not an isomorphism but in many interesting cases it is. If there is an open subset V of y such that the natural mapping

$$\psi_i(\mathcal{G}) : R^i f_* \mathcal{F}|_n \otimes \mathcal{G} \to R^i f_* (\mathcal{F} \otimes f^* \mathcal{G})$$

is an isomorphism for all quasi-coherent sheaves $\mathcal{G}$ on U we will say that $R^i f_* \mathcal{F}$ *commutes with base extension in* U. This implies in particular that $\psi_i(y)$ is an isomorphism for all y in U.

We begin our study with a result of Grauert which uses that our schemes are reduced.

Proposition 9.5.2. *If* $\dim_k H^i(X_y, \mathcal{F}_y)$ *is a constant* m *then* $R^i f_* \mathcal{F}$ *is locally free of rank* m *and commutes with base extension in* Y.

Proof. The problem is local on Y. We may return to the situation of the last proof. Thus rank$(\alpha^i|_y)$ and rank$(\alpha^{i-1}|_y)$ are constants in y. Thus α^i and α^{i-1} are homomorphisms of constant rank.

Lemma 9.5.3. *Let* $\alpha : \mathcal{F} \to \mathcal{G}$ *be a homomorphism between locally free coherent sheaves of constant rank. Then*

(a) Im α *is a locally free sheaf of constant rank which is a local direct summand of* $\mathcal{G}$ *and*

(b) Ker α *is a locally free sheaf of constant rank which is locally a direct summand on* $\mathcal{F}$.

Proof. Consider $\mathcal{H} = \operatorname{Cok} \alpha$. Then $\mathcal{H}|_y = \mathcal{G}|_y/\operatorname{Im} \mathcal{F}|_y$ for all y in Y. Thus dimension of $\mathcal{H}|_y$ is constant. Hence $\mathcal{H}$ is locally free of constant rank. Thus the exact sequence $0 \to \operatorname{Im} \alpha \to \mathcal{G} \to \mathcal{H} \to 0$ is locally split. So Im α is locally free of constant rank and a local direct summand of $\mathcal{G}$. The same reasoning applies to the exact sequence $0 \to \operatorname{Ker} \alpha \to \mathcal{F} \to \operatorname{Im} \alpha \to 0$.

Applying the lemma we have

$$0 \to \operatorname{Ker} \alpha^{i-1} \to K^{i-1} \to \operatorname{Im} \alpha^{i-1} \to 0,$$

$$0 \to \operatorname{Ker} \alpha^{i-1} \to K^i \to \operatorname{Im} \alpha^i \to 0$$

and

$$0 \to \operatorname{Im} \alpha^{i-1} \to \operatorname{Ker} \alpha^{i-1} \to R^i f_* \mathcal{F} \to 0$$

are all local split exact sequences of locally free sheaves of constant rank. In particular $R^i f_* \mathcal{F}$ is locally free. As the sequences are locally split they remain exact after tensoring with $\mathcal{G}$. It follows that $R^i f_*(\mathcal{F} \otimes f^* \mathcal{G}) \approx R^i f_*(\mathcal{F}) \otimes \mathcal{G}$; i.e. $R^i f_* \mathcal{F}$ commutes with base extension.

We will need another more general result.

Proposition 9.5.4. *Let y be a point of Y such that $H^i(X_y, \mathcal{F}_y) = 0$ for $i > p$. Then*

(a) *we may find a neighborhood U of y such that Proposition 9.4.1 is true with $n = p$,*

(b) $R^i f_* \mathcal{F} = 0$ *if $i \geq p$ and $R^i f_* \mathcal{F}$ commutes with base extension in U.*

Proof. For (a) we have the complex $0 \to K^0 \to K^1 \to \ldots \overset{\alpha^{n-1}}{\to} K^n \to 0$ which commutes with $R^* f_*(\mathcal{F} \otimes f^* \mathcal{G})$ in a neighborhood V by Proposition 9.5.1. If $n \leq p$ there is no problem. Assume that $n > p$. Then $\operatorname{Cok}(\alpha^{n-1}(y)) \approx H^n(X_y, \mathcal{F}_y)$ is zero. Replacing V by a smaller neighborhood we may assume that α^{n-1} is surjective. Thus we may write $\alpha^{n-1} : K^{n-1} \to K^n$ as the projection of K^{n-1} onto a direct summand K^n. Let $K_{\text{new}}^{n-1} = K^{n-1}/K_n$. Taking V still small we may assume that K_{new}^{n-1} is free. Consider the complex $K_{\text{new}}^n : 0 \to K^0 \to \ldots \to K^{n-2} \to K_{\text{new}}^{n-1} \to 0$. We have a mapping $\psi : K^* \to K_{\text{new}}^*$ which gives an isomorphism on cohomology. Hence by the mapping cone argument $\psi \otimes \mathcal{G}$ induces an isomorphism on cohomology for all $\mathcal{G}$. This proves (a) by induction on n.

For (b) $R^i f_* \mathcal{F}$ is the i-th homology of K^*. If $i > p$ then K^* is zero in this degree. Then $R^i f_* \mathcal{F} = 0$ if $i > p$. Now $R^0 f_*(\mathcal{F} \otimes f^* \mathcal{G})$ is isomorphic to $\operatorname{Cok}(\alpha^{p-1} \otimes \mathcal{G}) \approx \operatorname{Cok}(\alpha^{p-1}) \otimes \mathcal{G} \approx R^p f_* \mathcal{F} \otimes \mathcal{G}$. Thus $R^p f_* \mathcal{F}$ commutes with base extension in U.   □

There is another theorem of Grothendieck related to base extension.

Theorem 9.5.5. *There is a coherent sheaf Q together with a natural isomorphism $\psi_{\mathcal{G}} : \operatorname{Hom}(Q, \mathcal{G}) \overset{\to}{\approx} f_*(\mathcal{F} \otimes f^* \mathcal{G})$ for all quasi-coherent $\mathcal{O}_Y$-modules $\mathcal{G}$.*

Proof. By general nonsense Q is determined up to unique isomorphism. If we construct local solutions to the problem they will automatically patch together to give a global solution. Thus we may pass to the situation of Theorem 9.4.1 where we have the complex $0 \to K^0 \overset{\alpha^0}{\to} K^1$ such that $\mathrm{Ker}(\alpha^0 \otimes \mathcal{G}) \approx f_*(\mathcal{G} \otimes f^*\mathcal{F})$. Let $Q = \mathrm{Cok}(\alpha^{0\wedge} : K^{1\wedge} \to K^{0\wedge})$.

Then we have an exact sequence

$$
\begin{array}{ccccc}
Hom(K^{1\wedge}, \mathcal{G}) & \xleftarrow{\;Hom(\alpha^{0\wedge}, 1_{\mathcal{G}})\;} & Hom(K^{0\wedge}, \mathcal{G}) & \longleftarrow Hom(Q, \mathcal{G}) \longleftarrow 0 \\
\| & & \| & \\
K^1 \otimes \mathcal{G} & \xleftarrow{\quad \alpha^0 \otimes 1_{\mathcal{G}} \quad} & K^0 \otimes \mathcal{G} &
\end{array}
$$

Thus $Hom(Q, \mathcal{G}) \cong \mathrm{Ker}(\alpha^0 \otimes \mathcal{G}) \cong f_*(\mathcal{F} \otimes f^*\mathcal{G})$ and this is clearly natural in $\mathcal{G}$. $\qquad\square$

Exercise 9.5.6. Assume that $\psi_i(y)$ is surjective.

(a) $R^i f_* \mathcal{F}$ commutes with base extension near y.

(b) $R^i f_* \mathcal{F}$ is locally free near y iff $\psi_{i-1}(y)$ is surjective.

(Hint: choose $\ell_1, \ldots, \ell_n$ in $\mathrm{Ker}\,\alpha^i$ such that $\ell_1(y), \ldots, \ell_n(y)$ are a basis of $H^i(X_y, \mathcal{F}_y)$. Then take $m_1, \ldots, m_r$ in K^{n-1} such that $\alpha^{i-1}(m_1)(y), \ldots, \alpha^{i-1}(m_r)(y), \ell_1(y), \ldots, \ell_m(y)$ are a basis of $\mathrm{Ker}(\alpha^i(y))$. Then the obvious mapping $\psi : A = \oplus \mathcal{O}_u^{\otimes n} \otimes \mathcal{O}_u^{\otimes r} \to K^i$ gives an isomorphism between A and $\mathrm{Ker}(\alpha^i)$ in a neighborhood of y and $Im(\psi)$ is locally a direct summand of K^i.)

10

Applications

10.1 Embedding in projective space

We begin with

Lemma 10.1.1. *Let X be a closed subvariety of $Y \times Z$ where Y and Z are varieties and Z is projective. Let $f : X \to Y$ be the projection. Then f is finite if for all points y in Y the set $f^{-1}(y)$ is finite.*

Proof. We need to prove that f is an affine morphism because $f_* \mathcal{O}_X$ is coherent by Theorem 9.3.3. Fix y. We need to find an open affine neighborhood V such that $f^{-1}V$ is affine. We may assume that Y is affine. Now we may assume that Z is $\mathbf{P}^n$. Thus $f^{-1}(y) = \{y\} \times F$ where F is finite subset of $\mathbf{P}^n$. We can find a hyperplane $L = 0$ not meeting F. Then $F \subset D(L) = \mathbf{A}^n$. So $U = X \cap (Y \times D(L))$ is a closed subset of an affine variety $Y \times D(L)$. Thus U is affine. Let $W =$ complement of $f(X - U)$. As Z is complete, W is open. By construction W is a neighborhood of y and $f^{-1}W \subset U$. Let g be a regular function on Y such that $V = D(g) \subset W$. Then V is affine and $f^{-1}V = D_V(f^\star G)$ is affine. $\qquad\square$

Proposition 10.1.2. *Let $\sigma_0, \ldots, \sigma_n$ be sections of an invertible sheaf $\mathcal{L}$ on a projective variety X. Let V be the linear span of the σ_i in $\Gamma(X, \mathcal{L})$. Then the sections $\sigma_0, \ldots, \sigma_n$ define an embedding in $\mathbf{P}^n$ if*

(a)   for each point x of X, there is a section v in V such that $v(x) \neq 0$,

(b)   for each pair $x \neq y$ of distinct points of x there is a section v in V such that $v(x) \neq 0$ and $v(y) = 0$,

(c)   for each point of X the sections v in V such that $v(x) = 0$ have the properties that the $dv|_x$ span $m_x/m_x^2 \otimes_k \mathcal{L}|_x$ where $dv|_x$ is the image of v in $m_x \mathcal{L}/m_x^2 \mathcal{L} = m_x/m_x^2 \otimes_k \mathcal{L}|_x$.

Proof.   By Lemma 5.7.1, (a) implies that $(\sigma_0(x), \ldots, \sigma_n(x))$ define a morphism $g : X \to \mathbf{P}^n$. (b) means that the hyperplanes through $g(x)$ and $g(y)$ are distinct. Thus g is injective. We can factor $g : X \xrightarrow[\text{graph}]{} X \times$ $\mathbf{P}^n \xrightarrow[\text{projection}]{f} \mathbf{P}^n$ as $X \approx \text{graph}(g)$ and this is closed in $X \times \mathbf{P}^n$. By Lemma 10.1.1 g is a finite morphism. To show that g is an isomorphism we need to check that $\mathcal{O}_{\mathbf{P}^n}|_{g(x)} \to \mathcal{O}_X|_x$ is surjective for a point x of X. This means that $\Gamma(X, \mathcal{O}_X/(g^* m_{g(x)})$ is one dimensional. The sheaf in question is supported by x and has stalk $\mathcal{O}_{X,x}/g^* m_{g(x)}$. The assumption (b) means that $g^* m_{g(x)}$ spans m_x/m_x^2. Then by Nakayama's lemma $g^* m_{g(x)} = m_x$. Thus the stalk is k. $\square$

Let C be a complete smooth curve. Let $\mathcal{L}$ be an invertible sheaf on C. Let $\sigma_0, \ldots, \sigma_n$ be a basis of $\Gamma(C, \mathcal{L})$. Then $V = \Gamma(C, \mathcal{L})$. We want to know that we have a projective embedding of C.

Corollary 10.1.3.   *We have a projective embedding if $\Gamma(C, \mathcal{L}(-D))$ $\subset \Gamma(C, \mathcal{L})$ has codimension 2 for all effective divisors D of degree 2.*

Proof.   Straightforward. $\square$

Exercise 10.1.4.   If $\deg \mathcal{L} \geq 2$ genus$(C) + 1$ then $\mathcal{L}$ defines a projective embedding.

Exercise 10.1.5.   Let genus $(C) \geq 1, \Omega_C$ defines a projective embedding if genus$(C) \geq 2$ if C is not hyperelliptic.

10.2   Cohomological characterization of affine varieties

In this section we will prove a theorem of Serre.

Theorem 10.2.1.   *Let X be a variety. Then X is affine iff for each coherent sheaf $\mathcal{I}$ of ideals we have $H^1(X, \mathcal{I}) = 0$.*

Proof. The "only if" part is a special case of Theorem 8.2.2. For the "if", we have three steps.

Step 1. There are $f_1, \ldots, f_n$ in $k[X]$ such that the $D(f_i)$ are affine and cover X.

Step 2. There are $a_1, \ldots, a_n$ in $k[X]$ such that $1 = \sum a_i f_i$.

Step 3. The finale.

For Step 1 as X is quasi-compact we need only show that each point of y is contained in an affine $D(f)$ for f in $k[X]$. Let U be an affine neighborhood of x; we may find f such that $x \in D(f) \subset U$. Thus we want f to vanish on $Y = X - U$ and not at x. In other words f is a section of the ideal $\mathcal{I}_Y$ of Y such that the image in $\mathcal{O}_{\{x\}}$ is non-zero, say 1. Consider the exact sequence $0 \to \mathcal{I} \to \mathcal{I}_Y \to \mathcal{O}_{\{x\}} \to 0$. As $H^1(X, \mathcal{I}) = 0$ we can lift the section 1 of $\mathcal{O}_{\{x\}}$ to such an f.

For Step 2 consider the homomorphism $f : \bigoplus \mathcal{O}_X^{\oplus n} \to \mathcal{O}_X$ sending (a_i) to $\sum a_i f_i$. As the $D(f_i)$ cover X, f is surjective and we want to lift the section 1 to $\Gamma(X, \bigoplus \mathcal{O}_X^{\oplus n})$. Thus it is possible if $H^i(X, \mathrm{Ker}\, f) = 0$ but $\mathrm{Ker}(f)$ clearly has a coordinate filtration with coherent ideals as composition factors. By the long exact sequence of cohomologies, we have the required vanishing.

For Step 3 let $g_{i,1}, \ldots, g_{i,m_i}$ be elements of $k[X]$ such that they and $\frac{1}{f_i}$ generate $k[D(f_i)]$.

Let B be the subring of $k[X]$ to be generated by the f's, a's and g's. Then we have a morphism $g : X \to \mathrm{Spec}(B)$. By construction $D(f_i)$ cover $\mathrm{Spec}(B)$ and g is locally an isomorphism on the $D(f_i)$. Thus g is an isomorphism. Hence X is affine. $\qquad\square$

10.3 Computing the genus of a plane curve and Bezout's theorem

Let C be a smooth curve in $\mathbf{P}^2$. The degree of C = degree of an irreducible homogeneous equation of C. Then $\mathcal{O}_{\mathbf{P}^2}(-C) \approx \mathcal{O}_{\mathbf{P}^2}(-\deg C)$.

Lemma 10.3.1. *The genus g of C is $\frac{(d-1)(d-2)}{2}$ where $d = \deg C$.*

Proof. We have an exact sequence

$$0 \to \mathcal{O}_{\mathbf{P}^2}(-d) \to \mathcal{O}_{\mathbf{P}^2} \to \mathcal{O}_C \to 0.$$

Then $1 - g = \chi(\mathcal{O}_C) = \chi(\mathcal{O}_{\mathbf{P}^2}) - \chi(\mathcal{O}_{\mathbf{P}^2}(-d)) = 1 - \chi(\mathcal{O}_{\mathbf{P}^2}(-d))$. By Exercise 9.3.2 we have $g = \frac{(-d+2)(-d+1)}{2}$. The result follows. □

If C is singular then $\chi(\mathcal{O}_C) = 1 - p_a$ where p_a is the *arithmetic genus* of C. Thus even in the singular case the lemma computes p_a. The geometric genus p_g of C is the genus of $\tilde{C}$ where $\tilde{C}$ is the normalization of C in $k(C)$.

Exercise 10.3.2. $p_a - p_g = \sum\limits_{c \in C} \dim_k(f_* \mathcal{O}_{\tilde{C}}/\mathcal{O}_C)$ which is a finite sum where $f : \tilde{C} \to C$ is the canonical mapping.

Let C and D be two distinct plane curves in $\mathbf{P}^2$. Then the intersection $C \cap D$ is a finite number of points. If p is a point of $\mathbf{P}^2$ the *intersection multiplicity* $I(C, D : p)$ of C and D at p is $\dim_k \mathcal{O}_{\mathbf{P}^2,p}/(\mathcal{I}_{C,p} + \mathcal{I}_{D,p}.)$

Proposition 10.3.3. (Bezout)
$$\sum_{p \in \mathbf{P}^2} I(C, D : p) = \deg C \cdot \deg D.$$

Proof. Clearly the sum is $\dim_k \Gamma(\mathbf{P}^2, \mathcal{O}_{C \cap D}) = \chi(\mathcal{O}_{C \cap D})$ where
$$\mathcal{O}_{C \cap D} = \mathcal{O}_{\mathbf{P}^2}/(\mathcal{I}_C + \mathcal{I}_D.)$$
As the equations f and g of C and D are relatively prime we have an exact sequence
$$0 \to \mathcal{O}_{\mathbf{P}^2}(-\deg f - \deg g) \to \mathcal{O}_{\mathbf{P}^2}(-\deg f) + \mathcal{O}_{\mathbf{P}^2}(-\deg g) \to$$
$$\mathcal{O}_{\mathbf{P}^2} \to \mathcal{O}_{C \cap D} \to 0.$$
Thus $\chi(\mathcal{O}_{C \cap D}) = \chi(\mathcal{O}_{\mathbf{P}^2}) - \chi(\mathcal{O}_{\mathbf{P}^2}(-\deg f)) - \chi(\mathcal{O}_{\mathbf{P}^2}(-\deg g)) + \chi(\mathcal{O}_{\mathbf{P}^2}(-\deg f - \deg g))$. The result now follows from Exercise 9.3.2 by a simple calculation. □

We may give a more instructive proof that works when C is smooth (it always works in general but we have not developed the degree of an invertible sheaf on a singular curve). The proof is
$$\dim_k \Gamma(\mathbf{P}^2, \mathcal{O}_{C \cap D}) = \chi(\mathcal{O}_{C \cap D}) = \deg(\mathcal{O}_{\mathbf{P}^2}(D)|_C) =$$
$$\deg D \deg(\mathcal{O}_{\mathbf{P}^2}(1)|_C) = (\dim \Gamma(\mathbf{P}^2, \mathcal{O}_{C \cap L})) \deg D$$
where L is a line $\neq C$. Now exchanging C and L we get
$$\dim_k \Gamma(\mathbf{P}^2, \mathcal{O}_{C \cap D}) = \dim_k \Gamma(\mathbf{P}^2, \mathcal{O}_{L' \cap L}) \deg C \deg D$$
where L' and L are distinct lines. The last step is $\dim \Gamma(\mathbf{P}^2, \mathcal{O}_{L' \cap L}) = \dim_k k[\text{point}] = 1$.

10.4 Elliptic curves

An *elliptic curve* E is a smooth complete curve of genus one. The sheaf
of differentials Ω_E is trivial because it has degree $2 \cdot 1 - 2 = 0$ and
$\Gamma(E, \Omega_E)$ is one dimensional. By the Riemann–Roch theorem if $\mathcal{L}$ is
an invertible sheaf of positive degree then $\dim_k \Gamma(E, \mathcal{L}) = \deg \mathcal{L}$ and
$H^1(E, \mathcal{L}) = 0$.

We will classify all invertible sheaves of degree one. Consider $\text{Pic}_1(E)$
which is the set of isomorphism classes of invertible sheaves on E of
degree one.

Lemma 10.4.1. *The mapping $\smallint : E \to \text{Pic}_1(E)$ sending a point e of
E to the class of $\mathcal{O}_E(e)$ is a bijection.*

Proof. If $\mathcal{L}$ has degree one, then $\mathcal{L}$ has a non-zero section α. Thus
$\mathcal{L} \approx \mathcal{O}_E(D)$ where α corresponds to 1 and D is an effective divisor of
degree one; i.e. a point e. Thus $\mathcal{L} \approx \mathcal{O}_E(e)$. As α is unique up to a
constant, e determines $\mathcal{L}$ and conversely.   □

For the next proof we can explain better how to find e. Consider α as
a homomorphism $\mathcal{O}_E \to \mathcal{L}$; then $\hat{\alpha}: \mathcal{L}^{\otimes -1} \to \mathcal{O}_E$ has the ideal $\mathcal{O}_E(-e)$
as image.

Lemma 10.4.2. *Let $\mathcal{L}$ be an invertible sheaf on $E \times X$ where X is
a variety such that $\mathcal{L}_x \equiv \mathcal{L}|_{E \times x}$ has degree one for each point x of X.
Then the mapping $f : X \to E$ sending x to $\smallint^{-1}(\mathcal{L}_x)$ is a morphism.*

Proof. $\mathcal{L}$ is flat over X. As $H^1(E, \mathcal{L}_x) = 0$ for all x, the coherent
sheaf $\pi_{x*}\mathcal{L}$ is locally free and commutes with base extension. Thus
$\pi_{x*}\mathcal{L}$ is invertible as $H^0(E, \mathcal{L}_x)$ is one dimensional. We have a natural
homomorphism $\alpha : \pi_x^*(\pi_{x*}\mathcal{L}) \to \mathcal{L}$. Consider the ideal $\mathcal{I} = \text{image}(\tilde{\alpha} :
\pi_x^*(\pi_{x*}\mathcal{L}) \otimes \mathcal{L}^{\otimes -1} \to \mathcal{O}_{E \times X})$.

Claim. $Z = \text{zeroes}(\mathcal{I}) = \{(x, e) | f(x) = e\}$ and $\pi_X : Z \to X$ is an
isomorphism.

This of course will show that f is a morphism.

Let x be a point of X. Then $\pi_{x*}\mathcal{L}|_x \approx \Gamma(E, \mathcal{L}_x) \approx k$. Thus $\alpha|_{E \times x} :
\mathcal{O}_E \to \mathcal{L}_x$ is just the previous α. Thus $\mathcal{I}|_{E \times x} = \mathcal{O}_E(-e)$ where $\smallint e =
(\mathcal{L}|_x)$ so the first statement is clear. Now $\pi_X : Z \to X$ is one-to-one
and E is projective. Thus, by Lemma 10.1.1, $\pi_x : Z \to X$ is finite but
$\pi_{x*}(\mathcal{O}_Z)|_x = \Gamma(E, \mathcal{O}_e) = k$. Thus $\pi_{x*}\mathcal{O}_Z = \mathcal{O}_X$ and hence $\pi_X : Z \to X$
is an isomorphism.   □

Now let O be a fixed point of E. Then if $\mathrm{Pic}_0(E)$ is the isomorphism class of invertible sheaves of degree zero then the mapping $\int_0 :$ $E \to \mathrm{Pic}_0(E)$ which sends e to $\mathcal{O}_E(e-0)$ is a bijection. Clearly $\mathrm{Pic}_0(E)$ is a group under tensor product, thus E has a unique structure of abelian groups such that $\int_0$ is an isomorphism. Thus 0 is the zero of this group law. If e_1 and e_2 are points of E then e_1 sum e_2 is the point e_3 such that $\mathcal{O}_E(e_3 - 0) \approx \mathcal{O}_E(e_1 + e_2 - 2 \cdot 0)$. Thus $e_3 = \int^{-1} \mathcal{O}_E(e_1 + e_2 - 0)$. If e_2 is a point of E, then minus $e = f$ where $\mathcal{O}_E(f - 0) \approx \mathcal{O}_E(0 - e)$, thus $f = \int^{-1}(\mathcal{O}_E(2 \cdot 0 - e))$. We will next show

Proposition 10.4.3. *E is an algebraic group with this operation.*

Proof. We need to show that sum: $E \times E \to E$ and minus: $E \to E$ are morphisms. By Lemma 10.4.2 we just need to give invertible sheaves on $E \times (E \times E)$ and on $E \times E$ such that these maps are the corresponding morphisms f but this is easy. Consider $\mathcal{O}_{E \times E \times E}(\Delta_{1,2} + \Delta_{1,3} - 0 \times E \times E)$ where $\Delta_{i,j} = \{(e_1, e_2, e_3))|e_i = e_j\}$ and $\mathcal{O}_{E \times E}(2 \cdot 0 \times E - \Delta)$ these do the trick.　　□

10.5　Locally free coherent sheaves on $\mathbf{P}^1$

We will prove a result which was last proved by Grothendieck.

Proposition 10.5.1. *Let $\mathcal{F}$ be a locally free coherent sheaf on $\mathbf{P}^1$. Then $\mathcal{F} = \displaystyle\bigoplus_{1 \leq i \leq r} \mathcal{O}_{\mathbf{P}^1}(d_i)$ for some integers $d_1 \leq d_2 \leq \ldots \leq d_r$ which are uniquely determined by $\mathcal{F}$.*

Proof. Let us prove the uniqueness first. Let p be an integer. Let $\mathcal{F}_p = $ image of $\Gamma(\mathbf{P}^1, \mathcal{F}(-p)) \otimes_k \mathcal{O}_{\mathbf{P}^1}(+p)$ in $\mathcal{F}$. Then $\mathcal{F}_p = \displaystyle\bigoplus_{\substack{1 \leq i \leq r \\ d_i \geq p}} \mathcal{O}_{\mathbf{P}^1}(d_i)$. So rank $(\mathcal{F}_p / \mathcal{F}_{p+1}) = $ number of $d_i = p$. Thus $\mathcal{F}$ determines the d_i up to order.

As the proposition is the same for $\mathcal{F}$ and $\mathcal{F}(d)$, we may assume that $\Gamma(\mathbf{P}^1, \mathcal{F}) \neq 0$ and $\Gamma(\mathbf{P}^1, \mathcal{F}(-1)) = 0$. Let $\alpha \neq 0$ be a section of $\mathcal{F}$. Then we have a corresponding inclusion $\mathcal{O}_{\mathbf{P}^1} \subset \mathcal{F}$. The first point is that $\mathcal{F}/\mathcal{O}_{\mathbf{P}^1}$ is locally free. For otherwise $\mathcal{F}/\mathcal{O}_{\mathbf{P}^1}$ has torsion τ. Then the inverse image of τ in $\mathcal{F}$ is an invertible sheaf $\mathcal{L} \supset \mathcal{O}_{\mathbf{P}^1}$. If $\mathcal{L} \neq \mathcal{O}_{\mathbf{P}^1}$ then $\mathcal{L} \supset \mathcal{O}_{\mathbf{P}^1}(1)$. Hence $\mathcal{F}(-1)$ has a non-zero section which is impossible.

Thus we have the exact sequence

$(*)$ 　　　　　　　　　　$0 \to \mathcal{O}_{\mathbf{P}^1} \to \mathcal{F} \to \mathcal{F}' \to 0$

where $\mathcal{F}'$ is a locally free sheaf of one less rank than $\mathcal{F}'$. So by induction we know the proposition for $\mathcal{F}' \approx \bigoplus \mathcal{O}_{\mathbf{P}^1}(d_i)$. Now $\Gamma(\mathbf{P}^1, \mathcal{F}(-1)) = 0$ because we have the exact sequence

$$0 = \Gamma(\mathbf{P}^1, \mathcal{F}(-1)) \to \Gamma(\mathbf{P}^1, \mathcal{F}'(-1)) \overset{\delta}{\to} H^1(\mathbf{P}^1, \mathcal{O}_{\mathbf{P}}(-1)) = 0.$$

Therefore $d_i \leq 0$ for each i.

We next want to prove the sequence $(*)$ splits. Consider the dual sequence,

$$0 \to \mathcal{F}'^\wedge \to \mathcal{F}^\wedge \to \mathcal{O}_{\mathbf{P}^1} \to 0.$$

By now $\mathcal{F}'^\wedge = \bigoplus \mathcal{O}_{\mathbf{P}^1}(-d_i)$ where $d_i \leq 0$. Thus $H^1(\mathbf{P}^1, \mathcal{F}'^\wedge) = 0$. Therefore we may lift the section 1 of $\mathcal{O}_{\mathbf{P}^1}$ to a section of $\mathcal{F}^\wedge$. This gives a section $\mathcal{O}_{\mathbf{P}^1} \to \mathcal{F}^\wedge$ of $\mathcal{F}^\wedge \to \mathcal{O}_{\mathbf{P}^1} \to 0$. Thus the dual sequence splits so $(*)$ splits. $\qquad\qquad\square$

10.6 Regularity in codimension one

Let X be a smooth separated irreducible variety and Y be a complete variety. Let U be an open dense subset of X such that we have a morphism $f : U \to Y$. We will assume that U is taken to be a maximal open subset such that the rational mapping f is regular.

Proposition 10.6.1. $\dim X - U \leq \dim X - 2.$

Proof. We may assume that $f(X)$ contains a dense open subset of Y. Let $\pi : Z \to Y$ be a Chow covering where Z is projective and π is birational. We may factor f as $X \overset{g}{\to} Z \overset{\pi}{\to} Y$ where g is rational. Clearly it suffices to prove the proposition for g. Thus we may assume that Y is projective.

Let Γ be the closure of the graph of f in $X \times Y$. Then $\pi : \Gamma \to X$ is birational. Let $P = \{p \in \Gamma | \dim_p \pi^{-1}\pi p \geq 1\}$. Then P is closed by Proposition 6.4.5. Thus $\pi(P)$ is a proper closed subset of X. Clearly $U \subset X - \pi(P) = V$.

Claim. $U = X - \pi(P)$.

Here $\pi : \Gamma' \to V$ is finite by Lemma 10.1.1 where $\Gamma' = \pi^*V$. Thus $\pi_*\mathcal{O}_{\Gamma',x}$ is a finite $\mathcal{O}_{V,x}$ module for all x in V. Thus $\pi_*\mathcal{O}_{\Gamma',x}$ is integral over $\mathcal{O}_{V,x}$ but $\mathcal{O}_{V,x}$ is integrally closed as it is a UFD. Thus $\pi_*\mathcal{O}_{\Gamma',x} = \mathcal{O}_{V,x}$ or $\pi_*\mathcal{O}_{\Gamma'} = \mathcal{O}_V$. So $\pi : \Gamma' \to V$ is an isomorphism; i.e. f is regular on V. This proves the claim.

We need to show that dim $\pi(P) \leq$ dim $X - 2$. Now dim $X = $ dim $\Gamma >$ dim $P = $ dim $\pi(P) + 1$. Thus we are done. $\qquad\square$

10.7 One dimensional algebraic groups

Let G be an algebraic group where G is a curve. We want to prove

Proposition 10.7.1. *Either G is an elliptic curve, or*

$$G \approx \mathbb{G}_a \text{ or } \mathbb{G}_m.$$

Proof. The sheaf of differentials is trivial. One lets $T^*_{g^{-1}} \Omega_{G|g} \approx \Omega_{G|0}$ where T_g is translation by g. As this is clearly continuous in g, Ω_G is trivial.

If G is complete, genus$(G) = \frac{1}{2}$deg$(\Omega_G) + 1$. Thus genus$(G) = 1$ and hence G is elliptic.

Otherwise G is affine. Let $\tilde{G}$ be the smooth completion of G. Consider the rational mapping $\alpha : G \times \tilde{G} \mapsto \tilde{G}$ given by multiplication.

Claim. α is a morphism.

Let U be the maximal open subset where α is a morphism.

The complement of U is finite. Let $(h, \tilde{g})$ be one of the points of the complement. Then $\alpha(h, \tilde{g}) = (h \cdot \beta) \cdot (\beta^{-1}\tilde{g})$ for any β in G. If β is general then $(h \cdot \beta)(\beta^{-1}\tilde{g})$ is a morphism extending α. As translation by $\beta^{-1} : G \to G$ extends to a morphism $\tilde{G} \to \tilde{G}$ and $(h \cdot \beta, \beta^{-1}\tilde{g})$ is in U for general β, α extends to a morphism everywhere.

Let X be a tangent vector of the identity of G. Thus $\alpha_*(X \oplus 0)$ gives a vector field $\tilde{X}$ on $\tilde{G}$ which vanishes at the fixed points $\tilde{G} - G$. Thus $\Gamma(\tilde{G}, \Omega_{\tilde{G}}^{\otimes -1})$ is non-zero and has a section $\tilde{X}$ which vanishes on $\tilde{G} - G$. Therefore deg $\Omega_{\tilde{G}}^{\otimes -1} > 1$. So 2 genus$(\tilde{G}) - 2 < -1$. Thus genus$(\tilde{G}) = 0$ and $\tilde{X}$ vanishes twice at one point or once at two points. Therefore $\tilde{G} = \mathbb{P}^1$ and $G = \mathbb{A}^1$ or $\mathbb{A}^1 - \{0\}$.

Consider the case $G = \mathbb{A}^1$. We put the identity at 0. Now the multiplication $m : \mathbb{A}^1 \times \mathbb{A}^1 \to \mathbb{A}^1$ is given by $f(x,y) = f(x,y)$ where f is a polynomial in two variables. For fixed $x, y \mapsto f(x,y) : \mathbb{A}^1 \to \mathbb{A}^1$ is an isomorphism. Thus $f(x,y) = f_1(x)y + f_2(x)$ where $f_1(x)$ is never zero in $\mathbb{A}^1$. Hence $f(x,y) = Cy + f_2(x)$. Exchanging x and y we have $f(x,y) = Cy + Dx + E$ for constants C, D, E. Now $f(0,y) = y$ and $f(x,0) = x$. Thus $f(x,y) = x + y$ and G is $\mathbb{G}_a$.

Consider the case $G = \mathbb{A}^1 - \{0\}$. We put the identity at 1. Now m is given by $f(x,y)$ where f is polynomial in x, x^{-1} and y, y^{-1}. For fixed

$x, y \mapsto f(x,y)$ extends to an isomorphism $\mathbf{A}^1 \to \mathbf{A}^1$. Thus $f(x,y) = f_1(x)y + f_2(x)$ where $f_1(x)$ is never zero on $\mathbf{A}^1 - \{0\}$. Exchanging y and x we have $f(x,y) = Axy + Bx + Cy + D$ where $Ax + C$ is never zero on $\mathbf{A}^1 - \{0\}$ and $Ay + B$ is the same. Thus $C = B = 0$. $f(x,y) = Axy + D$ but $f(1,y) = y$ so $f(x,y) = xy$, and hence G is $\mathbf{G}_m$.   □

We have shown that given any point 0 of an elliptic curve E, E is an algebraic group with identity 0. We will next show that this group structure is unique.

Lemma 10.7.2. *If $\mu_1, \mu_2 : E \times E \to E$ are two algebraic group laws with identity 0 then $\mu_1 = \mu_2$.*

Proof. An easy proof of this relies on *Mumford's rigidity*.

Lemma 10.7.3. *Let $f : X \times Y \to Z$ be a morphism where X is irreducible and complete, Z is separated and Y is irreducible. If there is a point y_0 of Y such that $f(X \times y_0)$ is a point, then $f = g \circ \pi_Y$ where g is a morphism $g : Y \to Z$.*

Proof. Let W be an affine neighborhood of $f(X \times y_0)$. Then $F = \pi_Y(X \times Y - f^{-1}W)$ is a closed subset of Y as X is complete. By construction $V = Y - F$ is an open neighborhood of y_0. Let v be a point of V. Then f takes the complete variety $X \times y$ to the affine W. Thus $f(X \times y) = \{g(y)\}$ where $g(y)$ is a point. We next note that $g : V \to Z$ is a morphism as (*) $g(y) = f(x_0, y)$ for some fixed x_0 in X. As Y is irreducible V is dense in Y and the formula (*) defines g on all of y. We must have $g(y) = f(x,y)$ for all x and y.   □

To prove the first lemma just note the identity is a morphism

$$(E, \mu_1) \to (E, \mu_2).$$

Consider the morphism $\pi : E \times E \to E$ given by $\mu_2(e_1 \times e_2)(\mu_2(e_1) \mu_2(e_2))^{-1}$. We want to prove $\pi(e_1, e_2) = 0$. Now $\pi(0, e_2) = 0 = \pi(e_1, 0)$. Thus apply the lemma $\pi(e_1, e_2) = g(e_2) = 0$.   □

10.8 Correspondences

Let C and D be two smooth complete curves. A *correspondence* from C to D is an invertible sheaf $\mathcal{L}$ on $C \times D$. For instance if $f : C \to D$ is a morphism then $\mathcal{O}_{C \times D}(graph(f))$ is a correspondence.

A correspondence $\mathcal{L}$ is trivial if it has the form $\pi_C^* \mathcal{M} \otimes \pi_D^* \mathcal{N}$ where $\mathcal{M}$ and $\mathcal{N}$ are invertible sheaves on C and D. Two correspondences $\mathcal{L}_1$ and $\mathcal{L}_2$ are called *equivalent* if $\mathcal{L}_1 \otimes \mathcal{L}_2^{\otimes -1}$ is a trivial correspondence.

A correspondence $\mathcal{L}$ has two degrees. By Proposition 9.5.1(a), $\chi(\mathcal{L}|_{c \times D})$ is a constant in c. Let it be $\chi_D(\mathcal{L})$ and define $\deg_D(\mathcal{L})$ by the equation

$$\chi_D(\mathcal{L}) = \deg_D(\mathcal{L}) + 1 - \text{genus}(D).$$

Thus $\deg(\mathcal{L}_{c \times D}) = \deg_D(\mathcal{L})$ for all c in C. Reversing the roles of C and D we have $\deg_C(\mathcal{L})$ and $\chi_C(\mathcal{L})$ such that

$$\chi_C(\mathcal{L}) = \deg_C(\mathcal{L}) + 1 - \text{genus}(C) \text{ and } \deg(\mathcal{L}_{C \times d}) = \deg(\mathcal{L}) \text{ for all } d.$$

The numerical measure of the non-triviality of a correspondence $\mathcal{L}$ is the number $N(\mathcal{L}) \equiv -\chi(\mathcal{L}) + \chi_D(\mathcal{L})\chi_D(\mathcal{L})$. By the Künneth formula $N(\mathcal{L}) = 0$ if $\mathcal{L}$ is trivial but we will prove a much stronger fact.

Theorem 10.8.1.

(a) $N(\mathcal{L})$ *depends only on the correspondence class of* $\mathcal{L}$,

(b) $N(\mathcal{L}) \geq 0$,

(c) $N(\mathcal{L}) = 0$ *iff* $\mathcal{L}$ *is trivial.*

Proof. For (a) by symmetry it suffices to show that $N(\mathcal{L}) = N(\mathcal{L}(-c \times D))$ for all points c of C. We have an exact sequence

$$0 \to \mathcal{L}(-c \times D) \to \mathcal{L} \to \mathcal{L}|_{c \times D} \to 0.$$

Thus $\chi(\mathcal{L}) = \chi(\mathcal{L}(-c \times D)) + \chi_D(\mathcal{L})$ as $\chi_C(\mathcal{L}) = \chi_C(\mathcal{L}(-c \times D)) + 1$ and $\chi_D(\mathcal{L}) = \chi_D(\mathcal{L}(-c \times D))$. The result follows.

For (b) we may replace $\mathcal{L}$ by a suitably chosen representative $\mathcal{L}'$ of its correspondence class.

Claim. We can choose $\mathcal{L}'$ such that $\deg_C \mathcal{L}' = \text{genus}(C) - 1$ and $\Gamma(C, \mathcal{L}'|_{C \times e}) = 0$ for one point e of D.

Clearly if we take $\mathcal{L}' = \mathcal{L}(r(c_0 \times D))$ where c_0 is a point of C and $r = \text{genus}(C) - 1 - \deg_C(\mathcal{L})$ the first requirement is satisfied. Choose e arbitrarily; then if $\Gamma(C, \mathcal{L}'|_{C \times e}) = 0$ we are done. Otherwise replace $\mathcal{L}'$ by $\mathcal{L}'' = \mathcal{L}'(c_1 \times D - c_2 \times D)$ for general points c_1 and c_2 of C. Then $\dim_k \Gamma(C, \mathcal{L}''|_{C \times e}) = \dim_k \Gamma(C, \mathcal{L}'|_{C \times e}) - 1$. In fact take c_2 not to be a base point of $\Gamma(\mathcal{L}'|_{C \times e})$ and c_1 not to be a base point of $\Gamma(\Omega_C \otimes (\mathcal{L}'|_{C \times e})^{\otimes -1}(c_2))$ which is possible because $\chi_C(\mathcal{L}') = 0$. Thus $\Gamma(C, \mathcal{L}'|_{C \times e})$ decreases under the change and we are done by induction.

We next prove

Claim. $R^i\pi_{D*}\mathcal{L}' = 0$ unless $i = 1$ in which case it is a torsional coherent sheaf and $H^i(C \times D, \mathcal{L}') = 0$ unless $i = 1$ in which case it is isomorphic to $\Gamma(D, R^1\pi_{D*}\mathcal{L}')$.

We will first note how this implies (*b*). As $N(\mathcal{L}') = -\chi(\mathcal{L}') + 0 \cdot \star = -\chi(\mathcal{L}') =\dim H^1(C \times D, \mathcal{L}')$ we see that $N(\mathcal{L}') \geq 0$ and get a start on (*c*) by noting that $N(\mathcal{L}') = 0$ iff $R^i\pi_{D*}\mathcal{L}' = 0$ and consequently $N(\mathcal{L}') = 0$ if $\mathcal{L}'$ is trivial.

The first statement follows from base extension. As $\chi(\mathcal{L}|_{C\times e}) = 0$ we have $H^i(C, \mathcal{L}|_{C\times e}) = 0$ for all i. Therefore there is a neighborhood U of e such that $R^i\pi_{D*}\mathcal{L}|_U = 0$ for all i. Thus $\pi_{D*}\mathcal{L}$ is zero because it has no torsion and $R^1\pi_{D*}\mathcal{L}$ is a torsional coherent sheaf. Furthermore $R^i\pi_{D*}\mathcal{L} = 0$ for $i > 1$ by base extension because $H^i(C, \mathcal{L}|_{C\times d}) = 0$ for $i > 1$ and all d.

The second statement follows from the Leray spectral sequence which we don't have. So I will have to write the argument in detail. Let $\mathcal{L} \to D^*\mathcal{L}$ be the canonical flabby resolution of $\mathcal{L}$. Then $\pi_{D*}D^*\mathcal{L}$ is a complex of flabby sheaves with only one homology sheaves $R^1\pi_{D*}\mathcal{L}$ which is also flabby.

$$H^i(C \times D, \mathcal{L}) = i\text{-homology of } \Gamma(C \times D, D^*\mathcal{L})$$
$$= i\text{-homology of } \Gamma(D, \pi_{D*}D^*\mathcal{L}).$$

Thus we need only apply

Lemma 10.8.2. *If $0 \to \mathcal{F}^0 \to \mathcal{F}^1 \to \ldots$ is a complex of sheaves on a topological space X such that $H^i(X, \mathcal{F}^j) = 0$ for all j and $i > 0$ and the same for the homology sheaves $\mathcal{H}^j$ of the complex, then the i-homology of $\Gamma(X, \mathcal{F}^*)$ is naturally isomorphic to $\Gamma(X, \mathcal{H}^i)$.*

Proof. Break the complex up into short exact sequences $0 \to \mathcal{Z}^i \to \mathcal{F}^i \to \mathcal{B}^{i+1} \to 0$ and $0 \to \mathcal{B}^i \to \mathcal{Z}^i \to \mathcal{H}^i \to 0$. By ascending induction we prove that the higher cohomology of all these sheaves is zero. Hence applying $\Gamma(X,-)$ to them is exact. The result follows. $\square$

For (*c*) it remains to prove that $\mathcal{L}'$ is trivial if $R^i\pi_{D*}\mathcal{L}'$ is zero for all i. Let c be a point of C. Consider the exact sequence of sheaves

$$0 \to \mathcal{L}' \to \mathcal{L}'(c \times D) \to \mathcal{L}'(c \times D)|_{c\times D} \to 0.$$

By the long exact sequence of direct images $\pi_{D*}\mathcal{L}'(c \times D) \approx \pi_{D*}(\mathcal{L}'(c \times D)|_{c\times D})$ where the last sheaf say $\mathcal{M}$ is invertible. Thus we have a homomorphism $\mathcal{M} \hookrightarrow \pi_{D*}\mathcal{L}'(c \times D)$ and its adjoint $\psi : \pi_D^*\mathcal{M} \hookrightarrow \mathcal{L}'(c \times D)$ such that $\psi|_{c\times D} : \mathcal{M} \to \mathcal{L}'(c \times D)|_{c\times D}$ is an isomorphism. Consider the zero divisor Z of ψ. Then Z is an effective divisor such that $Z \cap c \times D$

is empty. Therefore an irreducible divisor R in Z is mapped by π_C to a closed subset of C which does not meet c. Thus the projection is a point r and $R = r \times D$. Therefore $\mathcal{L}'(c \times D) \approx \pi_D^* \mathcal{M}(\sum r_i \times D)$. Hence $\mathcal{L}'$ is trivial. □

Next we will show

Proposition 10.8.3. *N is a quadratic function on the group of correspondences; i.e. $< \mathcal{L}, \mathcal{N} >= N(\mathcal{L} \otimes \mathcal{N}) - N(\mathcal{L}) - N(\mathcal{N}) - N(\mathcal{O}_{C \times D})$ is biadditive and $N(\mathcal{L}) = \frac{1}{2} < \mathcal{L}, \mathcal{L} >$.*

This would be no problem if we knew the Riemann–Roch type theorem which computes $\chi(\mathcal{L})$ where $\mathcal{L}$ is an invertible sheaf on a smooth surface. I will give a direct proof.

Claim 1. $\mathcal{L}$ is equivalent to $\mathcal{O}_{C \times D}(-\sum R_i)$ where the R_i are disjoint smooth curves.

Note $E = c \times D + C \times d$ is ample on $C \times D$ for any c and d. (See Exercises 5.7.4 and 10.1.4). Then $\mathcal{L}^{\otimes -1} \sim \mathcal{L}^{\otimes -1}(mE)$ which is very ample. Thus we can take $\sum R_i$ to be a hyperplane section by Bertini's theorem.

Next we prove

Claim 2.
$$< \mathcal{L}, \mathcal{N} >= - \sum_i \deg(\mathcal{N}|_{R_i}) + \deg_C \mathcal{L} \deg_D \mathcal{N} + \deg_C \mathcal{N} \deg_D \mathcal{L}.$$

Consider the exact sequence
$$0 \to \mathcal{L} \otimes \mathcal{N} \to \mathcal{O}_{C \times D} \otimes \mathcal{N} \to \bigoplus_i \mathcal{N}|_{R_i} \to 0$$

$$0 \to \mathcal{L} \to \mathcal{O}_{C \times D} \to \bigoplus_i \mathcal{O}_{R_i} \to 0.$$

We get
$$\chi(\mathcal{L} \otimes \mathcal{N}) - \chi(\mathcal{N}) = \sum_i \chi(\mathcal{N}|_{R_i})$$

and
$$\chi(\mathcal{L}) \quad - \chi(\mathcal{O}_{C \times D}) = \sum_i \chi(\mathcal{O}_{R_i}).$$

Thus
$$-\chi(\mathcal{L} \otimes \mathcal{N}) + \chi(\mathcal{N}) + \chi(\mathcal{L}) - \chi(\mathcal{O}_{C \times D}) = \sum_i \chi(\mathcal{O}_{R_i}) - \chi(\mathcal{N}|_{P_i})$$

$$= - \sum_i \deg(\mathcal{N}|_{R_i}).$$

Now

$$\chi_C(\mathcal{L} \otimes \mathcal{N}) \cdot \chi_D(\mathcal{L} \otimes \mathcal{N}) - \chi_C(\mathcal{L})\chi_C(\mathcal{L})\,\chi_D(\mathcal{L})$$
$$- \chi_C(\mathcal{N})\chi_D(\mathcal{N}) - \chi_C(\mathcal{O}_{C\times D})$$
$$= \deg_C \mathcal{L} \cdot \deg_D \mathcal{N} + \deg_C \mathcal{N} \deg_D \mathcal{L}$$

by a calculation. Thus the lemma follows.

The claim shows that $< \mathcal{L}, \mathcal{N} >$ is additive in $\mathcal{N}$. Thus $< \mathcal{L}, \mathcal{N} >$ is biadditive by symmetry. It remains to show that $N(\mathcal{L}) = \frac{1}{2} < \mathcal{L}, \mathcal{L} >$. Now

$$\frac{1}{2} < \mathcal{L}, \mathcal{L} > = -\frac{1}{2} < \mathcal{L}, \mathcal{L}^{\otimes -1} >$$
$$= \frac{1}{2}(-N(\mathcal{L} \times \mathcal{L}^{\otimes -1}) + N(\mathcal{L}) + N(\mathcal{L}^{\otimes -1}) + N(\mathcal{O}_{C\times D}))$$
$$= \frac{1}{2}(N(\mathcal{L}) + N(\mathcal{L}^{\otimes -1}))$$

Therefore the remaining statement is equivalent to

Claim 3.　$N(\mathcal{L}) = N(\mathcal{L}^{\otimes -1})$.

As $\omega_{C\times D} = \pi_C^* \omega_C \otimes \pi_D^* \omega_D$ is a trivial correspondence we need to prove $N(\mathcal{L}^{\otimes -1}) = N(\omega_{C\times D} \otimes \mathcal{L})$ (This would follow from Serre duality for the surface $C \times D$.) Now we have exact sequence

$$0 \to \mathcal{L}^{\otimes -1} \to \mathcal{O}_{C\times D} \to \bigoplus_i \mathcal{O}|_{R_i} \to 0 \text{ and}$$

$$0 \to \omega_{C\times D} \to \omega_{C\times D}(\mathcal{L}) \to \bigoplus_i \omega_{R_i} \to 0$$

where the last step is the adjunction formula. Therefore

$$\chi(\mathcal{O}_{C\times D}) - \chi(\mathcal{L}^{\otimes -1}) = \sum_i \chi(\mathcal{O}_{R_i})$$

and

$$\chi(\omega_{C\times D} \otimes \mathcal{L}) - \chi(\omega_{C\times D}) = \sum_i \chi(\omega_{R_i})$$

which equal $-\sum_i \chi(\mathcal{O}_{R_i})$ by duality on the curves R_i. Thus

$$+\chi(\mathcal{O}_{C\times D}) - \chi(\mathcal{L}^{\otimes -1}) = \chi(\omega_{C\times D}) - \chi(\omega_{C\times D} \otimes \mathcal{L}).$$

Hence

$$-N(\mathcal{O}_{C\times D}) + N(\mathcal{L}^{\otimes -1}) = -N(\omega_{C\times D}) + N(\omega_{C\times D} \otimes \mathcal{L})$$

as

$$\chi_C(\mathcal{O}_{C\times D})\chi_D(\mathcal{O}_{C\times D}) - \chi_C(\mathcal{L}^{\otimes -1})\chi_D(\mathcal{L}^{\otimes -1}) =$$
$$\chi_C(\omega_{C\times D})\chi_D(\omega_{C\times D}) - \chi_C(\omega_{C\times D} \otimes \mathcal{L})\chi_D(\omega_{C\times D} \otimes \mathcal{L})$$

by duality again on C and D. Finally this gives

$$N(\mathcal{L}^{\otimes -1}) = N(\omega_{C \times D} \otimes \mathcal{L}).$$

$\square$

Corollary 10.8.4. $|<\mathcal{L}, \mathcal{M}>| \leq 2\sqrt{N(\mathcal{L})N(\mathcal{M})}.$

Proof. The standard inequality for a semi-positive quadratic form.

$\square$

We will apply this inequality when $\mathcal{L} = \mathcal{O}_{C \times D}(-\Gamma(f))$ and $\mathcal{M} = \mathcal{O}_{C \times D}(-\Gamma(g))$ where $f, g : C \to D$ are distinct morphisms. By definition $-\deg \mathcal{M}^{\otimes -1}|_{\Gamma(f)}$ is the number $< f, g >$ of points c such that $f(c) = g(c)$ counted with the right multiplicity. In fact $-\deg \mathcal{M}^{\otimes -1}|_{\Gamma(f)} = \dim_k \Gamma(C \times D, \mathcal{O}_{C \times D}/\mathcal{L}_{\Gamma(f)} + \mathcal{L}_{\Gamma(g)})$. In this case the inequality is a generalization of A. Weil's Riemann hypothesis for curves.

Theorem 10.8.5.

$$|\deg f + \deg g - < f, g >|$$

$$\leq 2\sqrt{[\text{genus}(D) + \text{genus}(C)(\deg f - 1)][\text{genus}(D) + \text{genus}(C)(\deg g - 1)]}$$

Proof. We need to show

(1) $< \mathcal{L}, \mathcal{M} >= \deg f + \deg g - < f, g >$ and
(2) $N(\mathcal{L}) = -$ genus$(D) -$ genus $(C)(\deg f - 1)$ and $N(\mathcal{M}) = -$ genus$(D) -$ genus$(C)(\deg g - 1)$.

Now $\deg_C \mathcal{O}_{C \times D}(\Gamma(f)) = 1$ and $\deg_D \mathcal{O}_{C \times D}(\Gamma(f)) = \deg f$. Thus (1) follows from the previous Claim 2. For (2) as the two statements are the same type we need to show

$$N(\mathcal{O}_{C \times D}(-\Gamma(f))) = -\text{genus}(D) - \text{genus}(C)(\deg f - 1).$$

We have the exact sequence

$$0 \to \mathcal{O}_{C \times D}(-\Gamma(f)) \to \mathcal{O}_{C \times D} \to \mathcal{O}_{\Gamma(f)} \to 0.$$

Thus

$$\chi(\mathcal{O}_{C \times D}(-\Gamma(f))) = \chi(\mathcal{O}_{C \times D}) - \chi(\mathcal{O}_C).$$

Now

$$N(\mathcal{O}_{C\times D}(-\Gamma(f))) = (\chi_C(\mathcal{O}_C)+1)(\chi_C(\mathcal{O}_D)+\deg f)$$
$$- \chi(\mathcal{O}_{C\times D}) - \chi(\mathcal{O}_C)$$
$$= (\chi_C(\mathcal{O}_C)\chi_D(\mathcal{O}_D) - \chi(\mathcal{O}_{C\times D}))$$
$$+ \chi_D(\mathcal{O}_D) + \chi_C(\mathcal{O}_C(\deg f + \deg f - \chi(\mathcal{O}_C))$$
$$= [\chi_D(\mathcal{O}_D) - \chi(\mathcal{O}_C)] - \operatorname{genus}(C)\deg f$$
$$= -\operatorname{genus}(D) + \operatorname{genus}(C) - \operatorname{genus}(C)\deg f$$
$$= -[\operatorname{genus}(D) + \operatorname{genus}(C)(\deg f - 1)].$$

$\square$

10.9 The Riemann-Roch Theorem for surfaces.

Let S be a smooth irreducible projective surface (dim $S = 2$). Let D and E be distinct closed curves on S. As in 10.3 we can define intersection number $I(C, D : p)$ at any point p of $C \cap D$ as $\dim_k \mathcal{O}_{S,p}/(\mathcal{I}_{C,p} + \mathcal{I}_{D,p})$. Let $\mathcal{I}(C; D) = \sum_{p \in C \cap D} I(C, D : p)$ be the total number of intersections. We want to define $I(C, D)$ using cohomology.

We have an exact sequence

$$0 \to \mathcal{O}_S(-C-D) \to \mathcal{O}_S(-C) \oplus \mathcal{O}_S(-D) \to \mathcal{O}_S \to \mathcal{O}_{C\cap D} \to 0.$$

As $C \cap D$ is only a finite number of points,

$$\Gamma(S, \mathcal{O}_{C\cap D}) = \bigoplus_{\mathcal{I} \in C \cap D} \mathcal{O}_{S,p}/(\mathcal{I}_{C,p} + \mathcal{I}_{D,p})$$

and $\mathcal{O}_{C\cap D}$ has no higher cohomology groups. Thus $\chi(\mathcal{O}_{C\cap D}) = I(C, D)$ but by the exact sequence

$$\chi(\mathcal{O}_{C\cap D}) = \chi(\mathcal{O}_S) - \chi(\mathcal{O}_S(-C)) - \chi(\mathcal{O}_S(-D)) + \chi(\mathcal{O}_S(-C-D)).$$

This motivates the following definition.

Let $\mathcal{L}$ and $\mathcal{M}$ be two invertible sheaves. The *intersection number* $[\mathcal{L}, \mathcal{M}]$ is by definition

$$\chi(\mathcal{O}_S) - \chi(\mathcal{L}^{\otimes-1}) - \chi(\mathcal{M}^{\otimes-1}) + \chi(\mathcal{L}^{\otimes-1} \times \mathcal{M}^{\otimes-1}).$$

Thus $[\mathcal{O}_S(C), \mathcal{O}_S(D)] = I(C, D)$ in the previous situation.

We have

Lemma 10.9.1. *Let $\mathcal{L}$ be an invertible sheaf on S. There exist disjoint smooth curves $C_1, \ldots, C_c$ on S and disjoint smooth curves $D_1, \ldots, D_d$ on S such that no D_i equals any C_j such that $\mathcal{L} \approx \mathcal{O}_S(\Sigma C_i - \Sigma D_j)$. (Actually one can choose c and d equal to one.)*

Proof. This is a consequence of Bertini Theorem. Let $\mathcal{N}$ be a very ample invertible sheaf. Choose d big enough so that $\mathcal{L} \otimes \mathcal{N}^{\otimes d}$ is generated by its sections. Thus $\mathcal{L} \otimes \mathcal{N}^{\otimes d+1}$ and $\mathcal{N}^{\otimes d+1}$ are very ample. Thus by Bertini Theorem for general sections of $\mathcal{L} \otimes \mathcal{N}^{\otimes d+1}$ and $\mathcal{N}^{\otimes d+1}$ their zeros C and D are smooth. Hence $\mathcal{L} \approx \mathcal{O}_s(C - D)$ and $C = \coprod C_i$ and $D = \coprod D_j$ as above. Q.E.D.

We may use this lemma to compute $[\mathcal{L}, \mathcal{M}]$ in

Lemma 10.9.2. *a)* $[\mathcal{L}, \mathcal{M}] = \sum \deg(\mathcal{M}|_{C_i}) - \sum \deg(\mathcal{M}|_{D_j})$
b) $[\mathcal{L}, \mathcal{M}]$ *is additive in* $\mathcal{M}$ *and hence* $\mathcal{L}$ *by symmetry.*

Proof. We have exact sequences
$$0 \to \mathcal{L} \to \mathcal{O}_S(C) \to \oplus_j \mathcal{O}_S(C)|_{D_j} \to 0 \text{ and}$$
$$0 \to \mathcal{L}^{\otimes -1} \to \mathcal{O}_S(D) \to \bigoplus_i \mathcal{O}_S(D)|_{C_i} \to 0$$

Thus
$$\chi(\mathcal{L}^{\otimes -1}) = \chi(\mathcal{O}_S(D)) - \sum \chi(\mathcal{O}_S(D)|_{C_i})$$
and
$$\chi(\mathcal{L}^{\otimes -1} \otimes \mathcal{M}^{\otimes -1}) = \chi(\mathcal{M}^{\otimes -1}(D)) - \sum \chi(\mathcal{M}^{\otimes -1}(D)|_{C_i}).$$

Hence
$$\begin{aligned}
[\mathcal{L}, \mathcal{M}] &= \chi(\mathcal{O}_S) - \chi(\mathcal{L}^{\otimes -1}) - \chi(\mathcal{M}^{\otimes -1}) + \chi(\mathcal{L}^{\otimes -1} \otimes \mathcal{M}^{\otimes -1}) \\
&= \chi(\mathcal{O}_S) - \chi(\mathcal{O}_S(-D)^{\otimes -1} - \chi(\mathcal{M}^{\otimes -1}) + \chi(\mathcal{M}(-D)^{\otimes -1}) \\
&\quad + \sum \chi(\mathcal{O}(D)|_{C_i} - \sum \chi(\mathcal{M}^{\otimes -1}(D))|_{C_i}) \\
&= [\mathcal{O}_S(-D), \mathcal{M}] + \sum \deg(\mathcal{M}|_{C_i}).
\end{aligned}$$

Now we have other exact sequences
$$0 \to \mathcal{O}_S \to \mathcal{O}_S(D) \to \mathcal{O}_S(D)|_D \to 0 \text{ and}$$
$$0 \to \mathcal{M}^{\otimes -1} \to \mathcal{M}^{\otimes -1}(D) \to \mathcal{M}^{\otimes -1}(D)|_D \to 0$$
$$[\mathcal{O}_S(-D), \mathcal{M}] = \chi(\mathcal{O}_S) - \chi(\mathcal{M}^{\otimes -1})$$
$$+\chi(\mathcal{M}^{\otimes -1}) - \sum \chi(\mathcal{O}_S(D_i)|_{D_i}) + \chi(\mathcal{M}^{-1}(D_i)|_{D_i}) = -\sum_j \deg(\mathcal{M}|_{D_j}).$$
$$\text{Q.E.D.}$$

Now we are ready for

Theorem 10.9.3 (Riemann-Roch). *Let $\mathcal{L}$ be an invertible sheaf on S. Then*
$$\chi(\mathcal{L}) = \frac{1}{2}[\mathcal{L}, \mathcal{L} \otimes \mathcal{O}_S \omega_S^{\otimes -1}] + \chi(\mathcal{O}_S).$$

Proof. Write $\mathcal{L} = \mathcal{O}_S(\sum C_i - \sum D_j)$ as before.

We have two exact sequences

$$0 \to \mathcal{L} \to \mathcal{O}_S(\textstyle\sum C_i) \to \bigoplus_j \mathcal{O}_S(\textstyle\sum C_i)|_{D_j} \to 0$$

and

$$0 \to \mathcal{O}_S \to \mathcal{O}_S(\textstyle\sum C_i) \to \bigoplus_k \mathcal{O}_S(\textstyle\sum C_i)|_{C_k} \to 0.$$

Thus

$$\begin{aligned}
\chi(\mathcal{L}) &= \chi(\mathcal{O}_S) + \textstyle\sum \chi(\mathcal{O}_S(\textstyle\sum C_i)|_{C_k} - \textstyle\sum \chi(\mathcal{O}_S(\textstyle\sum C_i)|_{D_j}) \\
&= \chi(\mathcal{O}_S) + \textstyle\sum_k \chi(\mathcal{O}_{C_k}) - \textstyle\sum_j \chi(\mathcal{O}_{D_j}) \\
&\quad + \textstyle\sum \deg(\mathcal{O}_S) + \textstyle\sum C_i|_{C_k}) - \deg(\mathcal{O}_S(\textstyle\sum C_i)|_{D_j}) \\
&= \chi(\mathcal{O}_S) + [\mathcal{O}_S(\textstyle\sum C_i), \mathcal{L}] + \textstyle\sum \chi(\mathcal{O}_{C_i}) - \textstyle\sum_j \chi(\mathcal{O}_{D_j}).
\end{aligned}$$

Hence we need to prove that

$$A : \sum_{ji} \chi(\mathcal{O}_{C_{ji}}) - \sum_j \chi(\mathcal{O}_{D_i}) =$$

$$\frac{-1}{2}[\omega_S, \mathcal{L}] - \frac{1}{2}[\mathcal{O}_S(\textstyle\sum C_i), \mathcal{L}] - \frac{1}{2}[\mathcal{O}_S(\textstyle\sum D_i), \mathcal{L}].$$

By the adjunction formula we have

$$\omega_{C_j} = \omega_S(\textstyle\sum C_i)|_{C_j} \text{ and } \omega_{D_i} = \omega_S(\textstyle\sum D_j)|_{D_i}.$$

Now if D is a smooth complete curve, $\chi(\mathcal{O}_E) = \frac{-1}{2}[\deg \Omega_C]$. Therefore the left side of A is

$$= -\tfrac{1}{2}[\textstyle\sum_j \deg(\omega_S(\textstyle\sum C_i)|_{C_i}] + \tfrac{1}{2}\textstyle\sum \deg(\omega_S(\textstyle\sum D_i|_{D_j})$$

$$= \tfrac{1}{2}[-[\omega_S, L] - [\mathcal{O}(\textstyle\sum C_i), \mathcal{O}(\textstyle\sum C_i)] + (\mathcal{O}(\textstyle\sum D_i), \mathcal{O}(\textstyle\sum D_i))]]$$

This equals the right side of A by the bilinearity of intersection.

Q.E.D.

Exercise 10.9.4 For a correspondence $\mathcal{L}$ on $C \times D$, shows that $N(\mathcal{L}) = \frac{-1}{2}[\mathcal{L}, \mathcal{L}] + \deg_C \mathcal{L} \deg_D \mathcal{L}$.

Appendix

A.1 Localization

Let A be a (commutative) ring with identity 1. A multiplicative subset S is a subset of A such that

(a) 1 is in S and
(b) if s_1 and s_2 are in S, then product $s_1 \cdot s_2$ is in S.

If S is a multiplicative subset of A, we can define a new ring A_S, the *localization* of A with respect to S. An element of A_S is an equivalence class of expressions a/s where a is in A and s is in S. The equivalence relation is

$$a/s \sim b/t \text{ iff } \exists u \in S \text{ such that } u(ta - sb) = 0 \text{ in } A.$$

One first checks that this is an equivalence relation. The addition in A_S is given by

$$a/s + b/t \sim ta + sb/st$$

and the multiplication in A_S is given by

$$a/s \cdot b/t \sim ab/st.$$

One next checks that these operations are well defined on equivalence classes and A_S is a commutative ring.

We have a natural homomorphism $\psi : A \to A_S$ which sends a to $a/1$.

Exercise A.1.1. Prove that Ker $\psi = \{a \in A | \exists s \in S \text{ such that } sa = 0\}$.

Exercise A.1.2. Let $\psi : A \to B$ be a homomorphism of rings.
Then we have a commutative diagram

iff $\psi(S)$ C contained units of B in which case $\overline{\psi}$ is uniquely determined by ψ.

Exercise A.1.3. Prove that as an A-algebra A_S is canonically isomorphic to $A[X_{s, s \in S}]/(sX_s - 1_{s \in S})$ where X_s are indeterminate.

Localization is a generalization of the usual construction of the quotient field $\mathbf{Q}$ of an integral domain $\mathbf{Z}$. In this case $Q = Z_{Z - \{0\}}$. We will give the most frequent examples of multiplicative sets.

Let f be an element of A. The set $(f) = \{1, f, f^2, \ldots\}$ is the minimal multiplicative set containing f. The ring

$$A_{(f)} = \{\frac{a}{f^i} \text{modulo} \ \frac{a}{f^i} \sim \frac{b}{f^j} \Leftrightarrow f^{j+n}a$$
$$= f^{i+n}b \text{ for some } n \geq 0\}$$

Let P be a prime ideal of A. Then $S = A - P$ is a multiplicative set. In this case

$$A_S \text{ is denoted by } A_P.$$

Exercise A.1.4. Show that A_P is a local ring with maximal ideal $\psi(P)A_P$.

Exercise A.1.5. $A_{(f)}$ is the zero ring iff f is nilpotent.

Given A and S we may localize an A-module M and obtain an A_S-module M_S. An element of M_S is an equivalence class of expressions m/s where m is in M and s is in S.
The equivalence relation is

$$m_1/s_1 \sim m_2/s_2 \text{ iff } t(s_2m_1 - s_1m_2) = 0 \text{ in } M$$

for some element t of S. The addition in M_S is $m_1/s_1 + m_2/s_2 = (s_2m_1 + s_1m_2/s_1s_2)$ and the scalar multiplication is $a/s \cdot m/t = am/st$. One may check that these definitions give an A_S-module M_S. We have a natural A-homomorphism $\psi : M \to M_S$ sending m to $m/1$. Thus we have a natural A_S-homomorphism $\alpha : M \otimes_A A_S \to M_S$.

Exercise A.1.6. α is an isomorphism.

Exercise A.1.7. Let $0 \to M_1 \to M_2 \to M_3 \to 0$ be an exact sequence of A-modules. Then we have an exact sequence

$$0 \to M_{1,S} \to M_{2,S} \to M_{3,S} \to 0 \text{ of } A_S\text{-modules.}$$

A.2 Direct limits

Let $(U, >)$ be a partially ordered set. Then $(U, >)$ is *directed* if for all u_1 and u_2 in U there is u_3 in U such that $u_1 \geq u_3$ and $u_2 \geq u_3$. We will assume that $(U, >)$ is directed.

A *direct system* $(F_u)_{u \in U}$ is a set F_u for each u in U together with a mapping $r_{u_2}^{u_1} : F_{u_1} \to F_{u_2}$ if $u_1 \geq u_2$ such that

(a) r_u^u = identity of F_u for all u and
(b) $r_{u_2}^{u_1} = r_{u_3}^{u_2} \circ r_{u_2}^{u_1}$ if $u_1 \geq u_2 \geq u_3$.

The *direct* limit $\varinjlim F_u$ of a direct system $(F_u)_{u \in U}$ is the set $\coprod_{u \in U} F_u$ modulo the relation that f_1 in F_{u_1} and f_2 in F_{u_2} are equivalent if there exists a u_3 such that $u_1 \geq u_3$ and $u_2 \geq u_3$ such that $r_{u_3}^{u_1}(f_1) = r_{u_3}^{u_2}(f_3)$. One checks easily that this is an equivalence relation. For each u in U we have a canonical mapping $s_u : F_u \to \varinjlim F_u$ sending f in F_u to the class represented by f. This mapping has the compatibility $s_{u_2} = r_{u_2}^{u_1} s_{u_1}$ if $u_1 \geq u_2$.

If $(F_u)_{u \in U}$ and $(G_u)_{u \in U}$ are two direct systems a *morphism* $\alpha : (F_u) \to (G_u)$ of direct systems is a mapping $\alpha_u : F_u \to G_u$ for all u in U such that $r_{u_2}^{u_1} \circ \alpha_{u_1} = \alpha_{u_2} \circ r_{u_2}^{u_1}$ when $u_1 \geq u_2$. In this case we have an induced mapping

$$\varinjlim \alpha : \varinjlim F_u \to \varinjlim G_u$$

which is determined by the equation $s_u \cdot \alpha_u = \varinjlim \alpha \cdot s_u$ for all u in U.

If $\mathcal{C}$ is a category, we may define a direct system $(F_u)_{u \in U}$ in $\mathcal{C}$. Here F_u are objects in $\mathcal{C}$ and the $r_{u_2}^{u_1}$ are morphisms in $\mathcal{C}$. For instance we may have a direct system of abelian groups, rings, modules, etc. Usually we can define a direct limit of $(F_u)_{u \in U}$ in $\mathcal{C}$. For instance if $\mathcal{C}$ is {abelian groups}, then $\varinjlim F_u$ is the set-theoretic direct limit with the addition: if f_1 is in F_{u_1} and f_2 is in F_{u_2} then

$$f_1 + f_2 \sim r_{u_3}^{u_1}(f_1) + r_{u_3}^{u_2}(f_2) \text{ where } u_1 \geq u_3 \text{ and } u_2 \geq u_3.$$

One simply checks that this addition is well-defined and gives $\varinjlim F_u$ the structure of abelian groups such that the s_u's are homomorphisms. Also one may define a morphism of direct systems in $\mathcal{C}$ by requiring that the α_u are morphisms in $\mathcal{C}$. If the direct limits exist in $\mathcal{C}$ as above then

$$\text{limit } \alpha : \varinjlim F_u \to \varinjlim G_u$$

is a morphism in $\mathcal{C}$. One important property of direct limits is exactness. An exact sequence of direct systems of abelian groups is the homomorphism of direct systems

$$0 \to (M_u)_{u \in U} \to (N_u)_{u \in U} \to (P_u)_{u \in U} \to 0$$

such that for all u the sequence $0 \to M_u \to N_u \to P_u \to 0$ is exact.

Exercise A.2.1. In this situation prove

$$0 \to \varinjlim M_u \to \varinjlim N_u \to \varinjlim P_u \to 0$$

is exact.

In this book I use the connection between localization and direct limits systematically. Let f be an element of a ring A and M be an A-module. Let $(U, >)$ be $\mathbf{N}$ with reverse ordering. Let $\frac{1}{f^n} \cdot M$ be F_n where $\frac{1}{f^n}$ is a bookkeeping symbol. Then we have a direct system $r_{n_2}^{n_1} : \frac{1}{f^{n_1}} M \to \frac{1}{f^{n_2}} M$ where $n_2 \geq n_1$ where $r_{n_2}^{n_1}(\frac{1}{f^{n_1}} m) = \frac{1}{f^{n_2}} \cdot f^{n_2 - n_1} m$.

Exercise A.2.1. Show that $\varinjlim \frac{1}{f^n} M = M_{(f)}$.

A.3 Eigenvectors

Let A be a set of linear operators on a k-vector space V. A vector v in V is an *eigenvector* for A if $a(v) = \lambda_a \cdot v$ for some k-valued function λ on A. Here λ is called the *eigenvalue* of v. For given λ the subset V_λ consisting of all eigenvectors v with eigenvalue λ is a subspace of V, called the λ-eigensubspace. We have a natural injection

$$\bigoplus_\lambda V_\lambda \hookrightarrow V$$

as non-zero eigenvectors with distinct eigenvalues are linearly independent by usual induction on the number of vectors. If this injection is an isomorphism we will say that V is spanned by eigenvectors.

A subspace W of V is A-*invariant* if $a(W) \leq W$ for all a in A.

Lemma A.3.1. *If W is an A-invariant subspace of V and V is spanned by eigenvectors then W is spanned by eigenvectors.*

Proof. Let w be an element. Then $w = \sum_\lambda v_\lambda$ where v_λ is a λ-eigenvector in V and the sum is finite. It will suffice to show that v_λ is in W for all λ. Now $a(w)$ is in W as $\lambda'_a \cdot w - a(w) = \sum(\lambda'_a - \lambda_a)v_\lambda$ for any λ' say one of the λ. Thus by induction on the number of λ,

$$(\lambda'_a - \lambda_a)v_\lambda \text{ is in } W \text{ for all } \lambda \text{ and } a.$$

As long as there are two distinct λ this implies that v_λ is in W. If there is only one λ then there is no problem. $\qquad\square$

Bibliography

[AB]. Auslander, M. and Buchsbaum, D. Homological dimension in local rings, *Trans. Amer. Math. Soc.*, **82**, 390–405 (1957).

[AM]. Atiyah, M. and MacDonald, I. *Introduction to Commutative Algebra*, Addison-Wesley Series in Mathematics, Addison-Wesley Publishing Company (1969).

[B]. Bourbaki, N. *Commutative Algebra*, chapters 1–7, Elements of mathematics, Springer-Verlag (1989).

[EGA]. Grothendieck, A. and Dieudonné, J. Eléments de Géométrie Algébrique:

EGA I. Le langage des schémas, *Publ. Math. IHES* **4** (1960).

EGA II. Etude globale élémentaire de quelques classes de morphismes, Ibid. **8** (1961).

EGA III. Etude cohomologique des faisceaux cohérents, Ibid. **11** and **17** (1963).

EGA IV. Etude locale des schémas et des morphismes de schémas, Ibid. **20** (1964), **24** (1965), **28** (1966), **32** (1967).

EGA I. Eléments de géométric algébrique, I. Springer-Verlag (1971).

[F]. Fulton, W. *Algebraic Curves*, Mathematics Lecture Note Series, The Benjamin/Cummings Publishing Company (1974).

[G1]. Godement, R. *Topologie algébrique et théorie des faisceaux*, Hermann (1958).

[G2-1]. Grothendieck, A. Sur quelques points d'algèbre homologique, *Tôhoku Math. J.* 9 (1957), 119–221.

[G2-2]. Grothendieck, A. *Local Cohomology*, Lecture Notes in Math. 41, Springer-Verlag (1967)

[H1]. Harthshorne, R. *Algebraic Geometry*, Springer-Verlag (1977).

[H2]. Hirzebruch, F. *Topological Methods in Algebraic Geometry*, Grundlehren 131, Springer-Verlag (1966).

[I]. Iitaka, S. *Algebraic Geometry, An Introduction to Birational Geometry of Algebraic Varieties*, Springer-Verlag (1982).

[K1-1]. Kempf, G. *Abelian Integrals*, Monografías de Instituto de Matemáticas, #13, Universidad Nacional Autónoma de Mexico, Mexico City (1984).

[K1-2]. Kempf, G. *Complex Abelian Varieties and Theta Functions*, Universitet, Springer-Verlag (1991).

[K2]. Kunz, E. *Introduction to Commutative Algebra and Algebraic Geometry*, Birkhäuser (1985).

[M1]. Matsumura, H. *Commutative Ring Theory*, Cambridge Studies in Advanced Mathematics 8, Cambridge Univ. Press (1989).

[M2-1]. Mumford, D. *The Red Book of Varieties and Schemes*, Lecture Notes in Mathematics 1358, Springer-Verlag (1988).

[M2-2]. Mumford, D. *Lectures on Curves on an Algebraic Surface*, Annals of Math. Studies 59, Princeton Univ. Press (1966).

[M2-3]. Mumford, D. *Curves and their Jacobians*, The Univ. of Michigan Press, Ann Arbor (1975).

[S1-1]. Serre, J.-P. Faisceaux algébriques cohérents, *Ann. of Math.* **61** (1955), 197–278.

[S1-2]. Serre, J.-P. *Algèbre locale multipliciative*, Troisième édition, Lecture Notes in Mathematics, Springer-Verlag (1989).

[S2]. Shafarevich, I. *Basic Algebraic Geometry*, Springer-Verlag, (1977).

[Z]. Zariski, O. *Algebraic Surfaces*, 2nd Suppl. ed., Ergebnisse 61, Springer-Verlag, (1971).

[Z-s]. Zariski, O. and Samuel, P. *Commutative Algebra*, vols I and II, Springer-Verlag (1960).

Glossary of notation

$A_{(a)}$, the localization of A at a 6
$[Aa : e^i A] = \{b \in A : e^i b \in Aa\}$ 74
$\mathbf{A}^1$, the affine line 3
$\mathbf{A}^n$, the affine n-space 10
$\mathbf{A}^n - \{0\}$, the punctured affine space 3
$\alpha(\sigma) \equiv \alpha(V)(\sigma)$ 39
$\alpha_x(\sigma_x) \equiv (\alpha(\sigma))_x$ 39
$\alpha_x(\mathcal{F})$ 43
$\mathcal{A}|_U$, the restriction of $\mathcal{A}$ to an open subset U 55
$\mathcal{A}(V) \equiv \bigcap_{v \in V}$ (integral closure in L of $\mathcal{O}_{Y,v}$) 82
$\mathcal{A}_X$-Hom$(f^*G, \mathcal{F})$ 66
$\mathcal{A}_Y$-Hom$(G, f_*\mathcal{F})$ 66

C^∞ 49
$\check{C}^*(\mathcal{F})$ 114
$\check{C}^*(\mathcal{F}) \otimes f^*(G)$ 118
$\chi(\mathcal{F})$, Euler characteristic of $\mathcal{F}$ 117
$\chi(\mathcal{O}_{\mathbf{P}^n}(r))$ 117
$(\text{Cok } \Gamma(X, f^*\psi))^\sim$ 67
$(\text{Cok } \psi)|_x$ 61
$\text{Cot}_x(X) \equiv m_x/m_x^2$, the Zariski cotangent space of X at x 70
$C(X)$, the cone in $\mathbf{A}^{n+1}$ over a closed set X in $\mathbf{P}^n$ 31
$\check{C}^*(X, \mathcal{F})$ 114

$D(\alpha)(\tau_v)_{v\in V} \equiv (\alpha_v(\tau_v))_{v\in V}$ 40

$D^\star(\mathcal{F})$ 98

$D(f)$ 1

$D(f)^\mathcal{F}$ 40

$D(\mathcal{F})(V) \equiv \prod_{v\in V} \mathcal{F}_v$ 43

$\deg \mathcal{L} \equiv \deg(D)$ for $\mathcal{L} \cong \mathcal{O}_C(D)$ 93

$\deg(\mathrm{div}(f))$ 93

$\deg f \equiv \dim_{k(D)} k(C)$ 93

$\deg (f^{-1}E)$ 93

$\det \mathcal{F} \equiv \Lambda^{\mathrm{rank}\ \mathcal{F}} \mathcal{F}$ 55

$df|_x \equiv$ the equivalence class of $f - f(x)$ in m_x/m_x^2 70

$d|_x : \mathcal{O}_{X,x} \to \mathrm{Cot}_x(X)$ 70

$d : \mathcal{O}_X \to \Omega_X$ 75

$\dim \mathbf{A}^n$ 20

$\dim f^{-1}(u)$ 80

$\dim f^{-1}(f(x))$ 81

$\dim_k H^i(X, \mathcal{F}_y)$ 121

$\dim_k \Gamma(C, \mathcal{F})$ 91

$\dim_k \Gamma(C, \mathcal{F}|_D)$ 92

$\dim_k(\mathcal{F}|_x)$ 61

$\dim_x(X) \equiv \max\{\dim C : C$ a component of X passing through $x\}$ 71

$\mathrm{div}(f)$ 93

divisor $f^{-1}E$ on C 93

divisor $f^{-1}d = \sum\limits_{c\in f^{-1}d} e_c \cdot c$ 93

$\mathrm{Div}(\mathcal{F}) \equiv \sum\limits_i \dim_k(\mathcal{F}_{c_i}) \cdot c_i$ 91

$\mathrm{Div}(\mathcal{F}|_D)$ 92

$\mathrm{Div}(X)$ 63

$\bigoplus_{i\in I} \mathcal{F}_i$ 53

$\bigoplus_{n\geq 0} \mathrm{Sym}^n(\mathrm{Cot}_x(X))$ 71

$D_r(\alpha) \equiv \{x \in X : \mathrm{rank}\ \alpha(x) \leq r\}$ 12

$d_X(f^*a) = f^*(d_Y a)$ 76

$E \equiv \pi_{\mathbf{A}^n}^{-1}(\{0\})$, the exceptional divisor at the origin 37

$f' : \Omega[Y] \otimes_{k[Y]} k[Y] \to \Omega[X]$ 76

$\mathcal{F}^\flat \equiv \mathrm{pre}\text{-}i(\mathcal{F})$ 44

$\mathcal{F}^\sharp$, the smallest subsheaf of $D(\mathcal{F})$ containing $\mathcal{F}^\flat$ 44

$\mathcal{F}'$, the extension of $\mathcal{F}$ by zero 59

$(f, f^*)\text{-}\mathrm{Hom}(G, \mathcal{F})$ 66

$f^{-1}G$ 66

f^*G 66

$(f_*\mathcal{O}_C|_U)\hat{\,}$, the dual of $f_*\mathcal{O}_C|_U$ 94

$f^* : \Omega_D \to \Omega_C$ 95

$f^*\Omega_D(\text{div } \Omega_{C/D})$ 95

$U^{\mathcal{F}}$ 57, 100

$\mathcal{F}(D) \equiv \cdot\mathcal{O}_C(D)$ 92

$\mathcal{F}|_D \equiv \mathcal{F}/\mathcal{F}(-D)$ 92

$\mathcal{F}_{(f)}$ 57

f-homomorphism 66

(f, f^*)-homomorphism 66

(f, f^*)-Hom$(G, \mathcal{F})$ 66

f-Hom$(G, \mathcal{F})$ 66

$\mathcal{F}(m) \equiv \mathcal{F} \otimes_{\mathcal{O}_x} \mathcal{O}_X(m)$ 60

$F_q : D \to D$ 95

$\mathcal{F} \otimes f^*$ 118

$\mathcal{F}_{\text{torsion}} \equiv \text{Ker}(\mathcal{F} \to \textbf{Rat}(\mathcal{F}))$ 91

$\mathcal{F}|_U$, the restriction of $\mathcal{F}$ on an open subset U of X 53

$\mathcal{F}_x$, stalk of $\mathcal{F}$ at x 39

$\mathcal{F}|_x \equiv \mathcal{F}_x/m_x\mathcal{F}_x$, a k-vector-space at a point x 60

$\mathbb{G}_a$, the additive algebraic group 31

$G(\mathcal{A}) \equiv \bigoplus_n m_x^n\mathcal{A}/m_x^{n+1}\mathcal{A}$ 79

$G/\mathcal{F} \equiv (\text{pre-}(G/\mathcal{F}))^{\sharp}$ 49

$\mathbb{GL}(n)$, the general linear algebraic group 31

$\mathbb{G}_m$, the multiplicative algebraic group 31

Graph(f) 28

$G|_X \equiv G/\mathcal{I}_X G$, the $\mathcal{O}_Y$-module annihilated by $\mathcal{I}_X$ 59

$G_x(\mathcal{O}_{X,x}) \equiv \bigoplus_{n\geq 0}(m_x^n/m_x^{n+1})$ 72

$\Gamma(C, \mathcal{L})$ 93

$\Gamma(D(f), \tilde{M})$ 56

$\Gamma(D(f), \mathcal{O}_X)$ 56

$\Gamma(U, -)$ 118

$\Gamma(X, D^*\mathcal{F})$ 98

$\Gamma(X, \tilde{M})$ 56

$\Gamma(X, \mathcal{O}_X)$ 56

$H^i(\mathbb{A}^n - \{0\}, \mathcal{O}_{\mathbb{A}^n})$ 113

$H^i(\mathbb{P}^n, \mathcal{O}_{\mathbb{P}^n}(d))$ 114

$H^1(C, \mathcal{F})$ 97

$H^i(X, \mathcal{F})$, the i-cohomology group of $\mathcal{F}$ 98

$H^i(X, \mathcal{F}(n))$ 116

$H^i(X, \mathcal{F} \otimes f^*(G))$ 118

$Hom_{\mathcal{A}}(\mathcal{N}, \mathcal{M})$ 55

$Hom(\mathcal{F}, \Omega_C)$, the global section of $Hom_{\mathcal{O}_C}(\mathcal{F}, \Omega_C)$ 107

$Hom_X(f^{-1}\mathcal{G}, \mathcal{F})$ 66

$Hom_Y(\mathcal{G}, f_*\mathcal{F})$ 66

$i : \mathcal{F} \to \mathbf{Rat}(\mathcal{F})$, a natural $\mathcal{O}_C$-linear mapping 91

$I(C, D : p)$, the intersection multiplicity of C and D at p 127

$IFI(X)$, the group of all inverible fractional ideals of Rat_X 63

$I_r(\alpha)$, the determinants of all $(r+1) \times (r+1)$ submatrices of α 12

$i(\sigma) \equiv (\sigma_u)_{u \in U}$ 40

$\mathcal{I}_X$, the ideal sheaf of regular functions on Y vanishing on X 59

$\mathcal{I}_X|_\Delta \equiv \mathcal{I}_X/\mathcal{I}_X^2$, the coherent $\mathcal{O}_Y$-module
annihilated by $\mathcal{I}_X$ 60

$K_U^V : \mathcal{G}(V)/\mathcal{F}(V) \to \mathcal{G}(U)/\mathcal{F}(U)$ 49

$k(X) \equiv \bigcup_{\emptyset \neq U \subseteq X \text{open}} k[U]$
the field of rational functions on X 62

$(k[X]_{(f)})\tilde{\ }$ 56

$\Lambda^n \mathcal{M}$, the exterior power 55

ℓ_y, a secant 73

$\varinjlim \mathcal{F}_i \equiv (\text{pre-}\varinjlim \mathcal{F}_i)^\sharp$ 51

Matrix $(f^* \psi_{ij})$ 67

$M \otimes_{A(X)} \mathcal{A}$, the associated sheaf 55

$\mathcal{M}\hat{\ } \equiv \mathcal{M}^{\otimes -1}$ 105

$M\tilde{\ } \equiv \pi_*(M\tilde{\ }]_{\mathbf{A}^{n+1}-\{0\} \times Y})_{\text{degree } 0}$ 117

$\tilde{\tilde{M}} \equiv M\tilde{\ }|_{C(X)-\{0\}}$ 61

$(M_{(f)})\tilde{\ }$ 56

$(M_{(f) \text{ degree } 0})\tilde{\ }$ 61

$\mathcal{M}|_{U'}$, the restriction of $\mathcal{A}$-module $\mathcal{M}$ to an open subset U 55

$M\tilde{\ }(U) \equiv (\tilde{\tilde{M}}(\pi^{-1}U)_{\text{degree } 0}$ 62

$(\tilde{M})_x$ 56

m_x/m_x^2 70

$\mathcal{N} \otimes_{\mathcal{A}} \mathcal{M}$ 55

$N(\mathcal{L}) \equiv -\chi(\mathfrak{L}) + \chi_D(\mathfrak{L})\chi_C(\mathfrak{L})$ 133

$\mathcal{O}_C(-\text{div}(f)) \equiv f \cdot \mathcal{O}_C$ 93

$\mathcal{O}_C(f^{-1}E) \equiv f^*(\mathcal{O}_D(E))$ 93

$\mathcal{O}_D \equiv \mathcal{O}_C|_D$ 92
$\mathcal{O}_{\mathbf{P}^n}(m)$ 60
$\mathcal{O}_{\mathbf{P}^n} = \mathcal{O}_{\mathbf{P}^n}(0)$ 60
$\mathcal{O}_{\mathbf{P}^n}(1)$ 65
$\mathcal{O}_X$, the structure sheaf 54
$\mathcal{O}_X(m) \equiv \mathcal{O}_{\mathbf{P}^n}(m)|_X \equiv \mathcal{O}_{\mathbf{P}^n}(m)/\mathcal{I}_X$ 60
$\mathcal{O}_X(D) \equiv \mathcal{I}_D^{-1}$ 63
$\mathcal{O}_X(U)$ 1
$\mathcal{O}_Y$-Alg-Hom$(\mathcal{A},\ g_*\mathcal{O}_Z)$ 82

Pic$(\mathbf{A}^n)$ 64
Pic$_0(E)$ 129
Pic$_1(E)$ 128
Pic$(\mathbf{P}^n)$ 65
Pic(X), the Picard group of X 62
pre-$\alpha(\mathcal{F})$ 43
pre-$(\mathcal{G}/\mathcal{F})$ 49
pre-$i(\mathcal{F})$ 44
pre-$\varinjlim (\mathcal{F}_i)(V) \equiv \varinjlim(\mathcal{F}_i(V))$ 51
Princ$_c(\mathcal{F}) \equiv$Rat$(\mathcal{F})/\mathcal{F}_c$ 96
Prin$(\mathcal{F}) \equiv \bigoplus_{c \in C}$ Prin $_c\mathcal{F}$ 96
Prin$(\mathcal{F})$, the sheaf of principal parts 96
$\mathbf{P}^1$, the projective line 3
$\mathbf{P}^n$, the projective n-sapce 10
$\mathbf{P}^{n^*}$, a projective space 83
$P(X)$, the principal fractional ideal in IFI(X) 63

Rad(A) 5
rank$(\alpha(x))$ 12
rank$\mathcal{F} \equiv$dim$_{k(X)}$Rat$(\mathcal{F})$ 90
Rat$(\mathcal{F})/\mathcal{F}_c$, the principal parts at c of rational sections of $\mathcal{F}$ 96
Rat$(\mathcal{F})$, the constant sheaf for a coherent $\mathcal{F}$ 90
Rat$_X$ 63
Res$_c(\omega)$, the residue of a rational differential ω at c 106
res$_U^V$ 38
res$_U^V$ in pre-$\varinjlim \mathcal{F}_i \equiv \varinjlim (res_U^V$ in $\mathcal{F}_i)$ for each open $U \subseteq V$ 51
res$_U^V((\tau_v)_{v \in V}) \equiv (\tau_u)_{u \in U}$ 40
$R^i f_*\mathcal{F}$, the higher direct image 102
$R^i f_*(\mathcal{F} \otimes f^*\mathcal{G})$ 118
$R^i f_*(\mathcal{F}_y)$, the skyscraper sheaf 120

$R^i \pi_{Y*} \mathcal{F}(n)$ 117

$\sigma_\alpha \equiv \sigma|_{V_\alpha}$ 41

$\sigma|_U$ 39

$\mathrm{Spec} A \equiv k\text{-Alg-Hom}(A, k)$ 4

$\mathrm{Spec}(A_{(f)})$ 5

$\mathrm{Spec}(A_{(f) \text{ degree } 0})$ 61

$\mathrm{Spec}(A/I)$ 9

$\mathrm{Supp}(\mathcal{F})$ 53

$\mathrm{Sym}(\mathrm{Cot}_z X)$, a polynomial ring 71

$\mathrm{Sym}^n \mathcal{M}$ 55

$(\mathrm{TC}_z X)_{\mathrm{red}} \equiv \mathrm{Spec}(G_z(\mathcal{O}_{X,z})/\sqrt{0})$, the tangent cone in $T_z X$ 73

$\mathrm{Tor}^1(L, \mathcal{O}_U)$ 120

$\mathrm{Tr}: k(C) \to k(D)$ 94

Tr' 111

$\mathrm{Tr}(f)(d)$ 95

$\mathrm{Tr}: f_*\mathcal{O}_C \to \mathcal{O}_{D'}$ an $\mathcal{O}_C$-linear mapping 94

$\mathrm{Tr}: f_*\Omega_C \to \Omega_D$ 111

$T_z X = \mathrm{Spec}(\bigoplus_{n \geq 0} \mathrm{Sym}^n(\mathrm{Cot}_z X))$ 6.3.71

$\psi \otimes_{k[Y]} G$ 119

$\omega_D \simeq \omega_X(D)|_D$, where D is a smooth divisor on X 79

$\omega_X \equiv \det \Omega_X$ 79

$\Omega[\mathbf{A}^n]$ 76

$\Omega_{C/D} \equiv \Omega_C/f_*\Omega_D$, a torsional $\mathcal{O}_C$-module 95

$\Omega_C \otimes_{\mathcal{O}_C} \mathcal{L}^{\otimes -1}$ 105

Ω_X, the sheaf of differentials on X 60

$\Omega_X|_z$ 77

$\Omega[X]$ 76

$\mathrm{zeroes}(\mathcal{I}) \equiv \mathrm{Support}(\mathcal{O}_X/\mathcal{I})$ 64

$< f, g >$ 137

$< \mathcal{L}, \mathcal{N} >$ 135

Index

A-invariant subspace 142
abelian sheaf 46
abelian variety 35
additive functor 98
adjoint 66, 134
adjunction formula 79
affine morphism 18
affine nullstellensatz 5
affine line 3
affine smooth curve 86
affine u-space 10
affine variety 3
algebraic group 30
ample invertible sheaves 68
arithmetic genus of C 127
ascending chain condition 62
associated sheaf $F^\natural$ 44
associated sheaf $M \otimes_{\mathcal{A}(X)} \mathcal{A}$ 55
associated sheaf $\mathrm{Sym}^n \mathcal{M}$ 55

basis for a topology 5, 7, 32
Bertini's theorem 83
Bezout's theorem 126
biadditive 135

birational morphism 34
blown up affine space 36

canonical quotient short exact sequences of flabby sheaves 47
categorical product 25
Cauchy residue 110
Čech cohomology 115
Chain condition 15
Characteristic of k is prime 94
Chow covering 89, 130
Chow's lemma 34
closed mapping 132
coherent module 58
coherent $\mathcal{O}_Y$-module $\mathcal{I}_X|_\Delta \equiv \mathcal{I}_X/\mathcal{I}_X^2$ annihilated by $\mathcal{I}_X$ 59
cohomology class 100
cohomology group $H^1(C,\mathcal{F})$ 97
cohomology group $H^i(X,\mathcal{F})$ 98
complete variety 33
complex $D^\star(\mathcal{F})$ 98
complex $\Gamma(X,\ D^\star\mathcal{F})$ 98
component of a topological space 17
cone 31
connected topological space 16
constant sheaf $\mathbf{Rat}(\mathcal{F})$ for a coherent $\mathcal{F}$ 90
coordinate filtration 126
correspondence 132
$\mathrm{Cot}_x(X) \equiv m_x/m_x^2$, the Zariski cotangent space of X at x 70
Cousin problem 96
Cramer's rule 6
curve $\equiv$ irreductible separated one-dimensional variety 85
cusp 75
$C(X)$ cone over a closed set X 31

decent presheaf 41
degree of a divisor 90
degree of an invertible sheaf on a smooth curve 93
degree of an invertible sheaf on a singular curve 93
dense 18
descending chain condition 16
determinants of all $(r+1) \times (r+1)$ submatrices of α 12
determinantal varieties 11

diagonal embedding 59
differential, $df|_x$ 70
differential form 72
differentials 49
dimension of a topological space 20
direct limit 50
  of sheaves 50
direct sum 54
direct system 141
  of sheaves 50
discontinuous sections 40
discrete valuation ring (DVR) 85
divisors 63

effective divisor 63
eigenvalues 142
eigenvectors 142
elliptic curve 35, 128
Euler characteristic $\chi(\mathcal{F})$ of $\mathcal{F}$ 117
Euler exact sequence 48
evaluation of a function at a point 8
exact sequence of abelian sheaves 47
exceptional divisor $E \equiv \pi_{\mathbf{A}^n}^{-1}(\{0\})$ at the origin 37
exterior power $\Lambda^n \mathcal{M}$ 55

f-homomorphism 66
(f, f^*)-homomorphism 66
field $k(X)$ of rational functions on irreducible X 62
filtration by coherent sheaves 91
finite morphism 18
finite surjective morphism 20
finitely generated torsional $\mathcal{O}_{C,c_i}$-modules 91
first order variation of a function at x 70
flabby resolution 102
flabby sheaves 46
flat module 118
flat sheaf over Y 118
fractional ideals 63
free abelian group 63
free $\mathcal{A}$-modules 55
function field of dimension one 88

genus of a curve 104
global invariant $\mathrm{Cok}(\alpha)$ of the sheaf $\mathcal{F}$ 97
global section Γ 94
global statements 93
$\mathbb{G}_m$-invariant 31
Godement's canonical flabby resolutions 46
graded module 61
graded ring 61
graph(f) 88
Grauert 121
Grothendieck 118

higher direct image $R^i f_* \mathcal{F}$ 102
Hilbert's nullstellensatz 5
Hilbert's basis theorem 15
Hirzebruch 117
Hirzebruch–Riemann–Roch theorem 117
homogeneous 11
homogeneous coordinate 28
homogeneous equation 126
homomorphism of abelian (pre-)sheaves 46
homotopically trivial 115
hyperelliptic curves 109
hyperplane 83
hypersurface 21

ideal sheaf 59
implicit differentiation 76
increasing sequence of ideals in the noetherian ring 86
induced structure of a space with functions 9
integral closure
    commutes with localization 83
    in quotient field 89
integral domain 16
intersection multiplicity $I(C,\ D:\ p)$ of C and D at p 127
intersection number 127
invertible $\mathcal{A}$-module 55
invertible ideals of $\mathcal{O}_X$ 64
invertible fractional ideals $\mathrm{IFI}(X)$ of $\mathbf{Rat}_X$ 63
invertible sheaves 62
irreducible divisors 63

irreducible topological space 16
isolated point = a component 20
isomorphism of spaces with functions 2
i-th cohomology group $H^i(X, \mathcal{F})$ 98

Künneth formula 114

Lagrange identity 111
left exact 67
Leray 38
Leray spectral sequence 134
limiting secant 73
linear change of variables 13
local coordinates 28
localization 6
locally closed 29
locally factorial 64
locally free $\mathcal{A}$-module 55
locally free coherent sheaf on a smooth curve 92
locally free $\mathcal{O}_D$-module of rank deg f 92
locally regular 1
locally trivial bundle of lines 37
locally vanishing principle 100
locus 32
long division 14
long exact sequence of homology groups 99
lower-semicontinuous 12
Lüroth 110

mapping cone 119
matrix of sections 55
matrix $(F^* \psi_{ij})$ 67
minimal prime ideal 17
Morphism
  of presheaves 38
  of spaces with functions 2
  of varieties 3
multiplicative subset 7
Mumford's rigidity 132

Nakayama's lemma 6

nilpotent 4
node 75
Noether 5
noetherian induction 52
noetherian local ring 74
noetherian topological space 15
Noether's normalization lemma 83
non-degenerate symmetric $k(D)$-bilinear form on $k(C)$ 94
normal variety 82
normalization of a variety X in $k(X)$ 82
nullstellensatz 5

obstruction to solving the Cousin problem 104
open affine subvarieties 11
open subspace of a space with functions 2
ordinary sheaf 97

parabola 75
partially ordered set 39
patching condition 41
Picard group of X, $\text{Pic}(X)$ 62
plane curves 127
Preparation lemma 13
presentation of $\mathcal{M}$ 58
presheaf 38
principal parts at c of rational sections of $\mathcal{F}$ 96
principal fractional ideals $P(X)$ of $\mathbf{Rat}_X$ 63
principal ideal theorem 22
principal part of ω at c 110
products 25
projective embedding 125
projective line $\mathbf{P}^1$ 3
projective n-space $\mathbf{P}^n$ 11
projective nullstellensatz 32
projective space $\mathbf{P}^{n^*}$ 83
projective subvariety 31
Proof of Lemma 1.4.1 14
Proof of Lemma 1.5.6 15
punctured affine space $\mathbf{A}^n - \{0\}$ 10
pure dimension 79
purely inseparable morphism 94

quadratic function on the group of correspondence 135
quasi-affine variety 10
quasi-coherent 55
quasi-compact 7
quasi-projective variety 11
quotient space with functions 10

radical Rad(A) of A 5
radical vector field 79
ramified at c 94
ramified morphism 94
ramification index 94
rank 55
rank, finite 55
rational differential 110
rational mapping 89
reduction to the diagonal 33
regular function 1
regular rational mapping 130
residue $\mathrm{Res}_c(\omega)$ of a rational differential ω at c 110
restriction $\mathcal{A}|_U$ of $\mathcal{A}$ to U 55
restriction $\mathcal{M}|_U$ of $\mathcal{A}$-module $\mathcal{M}$ to U 55
Riemann–Roch theorem 98
Riemann–Hurwitz theorem 109
right exact 67

secant ℓ_y 73
section 39
Segre embedding theorem 27
semi-positive quadratic forms 137
separable morphism 94
separated varieties 29
Serre 101
Serre duality 107
Serre vanishing theorem 125
sheaf 41
sheaf $\mathcal{A}$ of rings 54
sheaf $\mathbf{Prin}(\mathcal{F})$ of principal parts 96
sheaf of $\mathcal{A}$-modules 54
sheaf of C^∞ functions 49
sheaf of differentials 60

sheaf of fractional ideals $\mathcal{I}$ of $\mathbf{Rat}_X$ 63
sheaf $\mathcal{O}_X$ of regular functions on X 54
sheafification 44
sheafification ‖ 44
short exact sequence of abelian sheaves 47
singular curve 127
singular point 84
skyscraper sheaf $R^i f_*(\mathcal{F}_y)$ 120
smooth at a point 79
smooth complete curve 92
smooth completion of G 131
smooth curve 86
smooth divisor 79
smooth projective curve 88
smooth variety 71
snake homomorphism 120
space with functions 1
stalk F_x 39
stalk exactness 47
structure sheaf θ_X 43
sub-(pre-)sheaf 42
subvariety 9
support of $\mathcal{F}$ 53
syzygies 65

tame ramification point of f 95
tangent cone, $(\mathrm{TC}_x X)_{\mathrm{red}}$ 73
tangent space 71
tangent vector 131
torsional coherent sheaf 91
torsional element 91
torsional sheaf 91
torsion-free sheaf 91
translation by g 69
twisted sheaf $\mathcal{F}(m)$ 60

unique factorization domain (UFD) 21
unramified morphism 94
upper-semicontinuous 61

variational cohomology of a constant family of sheaves 103

variety (algebraic) 3
vector field 79
vector space $\mathcal{F}|_x$ at a point x 60
very ample invertible sheaves 68

Weil divisors 63
Weil–Riemann–Roch theorem 108
Weil's Riemann hypothesis 137

Zariski cotangent space 70
zero divisor of ψ 134

Printed in the United States
By Bookmasters